DEGRADABLE POLYMERS, RECYCLING, AND PLASTICS WASTE MANAGEMENT

PLASTICS ENGINEERING

Founding Editor

Donald E. Hudgin

Professor
Clemson University
Clemson, South Carolina

Additional Volumes in Preparation

DEGRADABLE POLYMERS, RECYCLING, AND PLASTICS WASTE MANAGEMENT

EDITED BY

ANN-CHRISTINE ALBERTSSON
Royal Institute of Technology
Stockholm, Sweden

SAMUEL J. HUANG
University of Connecticut
Storrs, Connecticut

 CRC Press
Taylor & Francis Group
Boca Raton London New York

CRC Press is an imprint of the
Taylor & Francis Group, an **informa** business

The contents of this volume were originally published in *Journal of Macromolecular Science, Part A—Pure and Applied Chemistry*, Volume A32, Number 4, 1995.

CRC Press
Taylor & Francis Group
6000 Broken Sound Parkway NW, Suite 300
Boca Raton, FL 33487-2742

First issued in paperback 2019

© 1995 by Taylor & Francis Group, LLC
CRC Press is an imprint of Taylor & Francis Group, an Informa business

No claim to original U.S. Government works

ISBN-13: 978-0-8247-9668-6 (hbk)
ISBN-13: 978-0-367-40166-5 (pbk)

Visit the Taylor & Francis Web site at
http://www.taylorandfrancis.com

and the CRC Press Web site at
http://www.crcpress.com

PREFACE

As the chairs for the Degradable Polymers working party of the IUPAC commission on functional polymers within the Macromolecular Division Committee, we organized an international workshop in April 1994 in Stockholm, Sweden. The aim of this workshop—Controlled Life-Cycle of Polymeric Materials: Biodegradable Polymers and Recycling—was to bring together scientists with an interest in design, synthesis, characterization, and long-term properties of degradable polymers.

The fate of polymers in nature seems to present a paradoxical problem at times; most polymers are designed and manufactured to resist environmental degradation (photodegradation, hydrolysis, oxidation, biodegradation, etc.) but are used for protective and/or structural purposes. Increasing awareness of solid waste management problems has led to the demand for polymers that do not have a harmful impact on the environment during any part of their life cycles. The complementary recycling of carbon, hydrogen, and nitrogen elements as well as of energy through reprocessing and biological and chemical conversion should be taken advantage of in polymer design. Research efforts in the 1970s by the editors and their colleagues on structure–property correlation and the mechanisms of degradation of polyethylene and step-growth polymers showed that the most important issue is not the type of degradation process, but rather how to develop tailor-made polymers with controllable lifetimes, taking into consideration environmentally acceptable manufacturing, application, recycling, and disposal methods. The ultimate goal is to obtain materials with an economically feasible degradation rate based on principles of the recycling of elements within the biosphere.

The accessibility of a polymer to degradative attack by microorganisms and higher biosystems is not solely related to its biological origin, but rather depends on its chemical architecture, composition, and processing. Both natural and synthetic macromolecules are degraded by microbial systems through hydrolysis and oxidation. Even polyethylene, which is nonhydrolyzable and in the lowest oxidation state, can be totally mineralized, albeit at a very slow rate.

The present book contains the contributions of 26 world-leading scientists who gave oral and poster presentations at the workshop. The workshop focused on the following topics: polymer waste management, polymers from renewable resources, processing and products, degradation and test methods, environmental aspects, future materials, and governmental policy for degradable polymers. The program was organized in seven sessions and each session was chaired by one university scientist and one person from industry. The 150 conference participants came from all over the world.

The main sponsor of the workshop was the Swedish Waste Research Council (AFR), which made the conference possible, and the chairman of the board of this council, C. Heinegård, director of the National Board of Industrial and Technical Development (NUTEK), also acted as chairperson of the panel discussion. We are very grateful for the strong interest shown by Swedish industry, which helped sponsor the workshop and sent many representatives to Stockholm. Special thanks go to Duni-Bilå, Eka Nobel, Lyckeby Stärkelsen AB, MoDo, Mölnlycke AB, and TetraPak Material AB. Finally, we want to thank all the contributors for their very interesting presentations and subsequent papers.

Ann-Christine Albertsson
Samuel J. Huang

CONTENTS

CONTRIBUTORS

Hideki Abe Polymer Chemistry Laboratory, The Institute of Physical and Chemical Research (RIKEN), Wako-shi, Saitama, Japan

Grażyna Adamus Institute of Polymer Chemistry, Polish Academy of Sciences, Zabrze, Poland

Ann-Christine Albertsson Department of Polymer Technology, Royal Institute of Technology, Stockholm, Sweden

Sahar Al-Malaika Polymer Processing and Performance Group, Department of Chemical Engineering and Applied Chemistry, Aston University, Birmingham, England

L. Ambrosio Department of Materials and Production Engineering and Institute of Composite Materials Technology, University of Naples "Federico II," Naples, Italy

G. Antranikian Department of Technical Microbiology, Institute of Biotechnology, Technical University Hamburg-Harburg, Hamburg, Germany

René Arnaud Laboratoire de Photochimie, URA CNRS 433, Universite Blaise Pascal, Clermont Ferrand, and Laboratoire de Photochimie Moléculaire et Macromoléculaire, Ensemble Universitaire des Cézeaux, Aubiere, France

C. Bastioli Novamont, Novara, Italy

Charles M. Buchanan   Research Laboratories, Eastman Chemical Company, Kingsport, Tennessee

F. Canganella   Department of Technical Microbiology, Institute of Biotechnology, Technical University Hamburg-Harburg, Hamburg, Germany

Sukhinder Chohan   Polymer Processing and Performance Group, Department of Chemical Engineering & Applied Chemistry, Aston University, Birmingham, England

M. Coker   Polymer Processing and Performance Group, Department of Chemical Engineering & Applied Chemistry, Aston University, Birmingham, England

J. Coudane   University Montpellier 1, CRBA-URA CNRS 1465, Faculty of Pharmacy, Montpellier, France

M. K. Cox   Biopolymers Group, ZENECA Bio Products, Billingham, Cleveland, England

Philippe Dabin   Laboratoire de Photochimie, URA CNRS 433, Universite Blaise Pascal, Clermont Ferrand, and Laboratoire de Photochimie Moléculaire et Macromoléculaire, Ensemble Universitaire des Cézeaux, Aubiere, France

A. Daro   Chimie des Polymères et Systèmes Organisés, Université Libre de Bruxelles, Bruxelles, Belgium

C. David   Chimie des Polymères et Systèmes Organisés, Université Libre de Bruxelles, Bruxelles, Belgium

W.-D. Deckwer   Biochemical Engineering, GBF, Gesellschaft für Biotechnologische Forschung mbH, Braunschweig, Germany

F. Degli Innocenti   Novamont, Novara, Italy

Yoshiharu Doi   Polymer Chemistry Laboratory, The Institute of Physical and Chemical Research (RIKEN), Wako-shi, Saitama, Japan

Debra D. Dorschel   Research Laboratories, Eastman Chemical Company, Kingsport, Tennessee

David Eberiel   Department of Biology, NSF Biodegradable Polymer Research Center, University of Massachusetts Lowell, Lowell, Massachusetts

Glenn R. Elion   International Communications and Energy, Chatham, Massachusetts

Annie Fauve   Laboratoire de Chimie Organique et Biologique, URA CNRS 433, Universite Blaise Pascal, Clermont Ferrand, and Ensemble Universitaire des Cézeaux, Aubiere, France

R. Clinton Fuller   Department of Biochemistry, University of Massachusetts, Amherst, Massachusetts

Robert M. Gardner   Research Laboratories, Eastman Chemical Company, Kingsport, Tennessee

Carl Grenthe   Department of Biochemistry, The Arrhenius Laboratories, Stockholm University, Stockholm, Sweden

Richard A. Gross   Department of Chemistry, NSF Biodegradable Polymer Research Center, University of Massachusetts Lowell, Lowell, Massachusetts

Ji-Dong Gu*   Department of Chemistry, NSF Biodegradable Polymer Research Center, University of Massachusetts Lowell, Lowell, Massachusetts

I. Guanella   Novamont, Novara, Italy

J. E. Guillet   Department of Chemistry, University of Toronto, Toronto, Ontario, Canada

Mika Härkönen   Department of Chemical Engineering, Helsinki University of Technology, Espoo, Finland

Hyoe Hatakeyama   National Institute of Materials and Chemical Research, Tsukuba, Ibaraki, Japan

Tatsuko Hatakeyama   National Institute of Materials and Chemical Research, Tsukuba, Ibaraki, Japan

Kari Hiltunen   Department of Chemical Engineering, Helsinki University of Technology, Espoo, Finland

Shigeo Hirose   National Institute of Materials and Chemical Research, Tsukuba, Ibaraki, Japan

Philippa J. Hocking   Department of Chemistry, McGill University, Montreal, Quebec, Canada

Samuel J. Huang   Polymer Science Program, Institute of Materials Science, University of Connecticut, Storrs, Connecticut

H. X. Huber   Department of Chemistry, University of Toronto, Toronto, Ontario, Canada

---

Current affiliation: Division of Applied Sciences, Harvard University, Cambridge, Massachusetts.

S. Iannace   Department of Materials and Production Engineering and Institute of Composite Materials Technology, University of Naples "Federico II," Naples, Italy

Merja Itävaara   VTT Biotechnology and Food Research, VTT, Finland

J. Jane   Center for Crops Utilization Research and Department of Food Science and Human Nutrition, Iowa State University, Ames, Iowa

Christer Jansson   Department of Biochemistry, The Arrhenius Laboratories, Stockholm University, Stockholm, Sweden

Zbigniew Jedliński   Institute of Polymer Chemistry, Polish Academy of Sciences, Zabrze, Poland

David L. Kaplan   US Army Natick RD&E Center, Natick, Massachusetts

Sigbritt Karlsson   Department of Polymer Technology, Royal Institute of Technology, Stockholm, Sweden

Fusako Kawai   Department of Biology, Kobe University of Commerce, Nishi-ku, Kobe, Japan

R. Kawamura   Central Laboratory, Rengo Co Ltd., Ohhiraki, Osaka, Japan

Joan Kelley   International Mycological Institute, Egham, Surrey, England

L. W. King   The Procter & Gamble Company, Cincinnati, Ohio

Dieter Klemm   Institute of Organic and Macromolecular Chemistry, University of Jena, Jena, Germany

M. Klingeberg   Department of Technical Microbiology, Institute of Biotechnology, Technical University Hamburg-Harburg, Hamburg, Germany

Ken Kobashigawa   Tropical Technology Center Ltd., Gushikawa, Okinawa, Japan

Fumitake Koizumi*   Polymer Chemistry Laboratory, The Institute of Physical and Chemical Research (RIKEN), Wako-shi, Saitama, Japan

Ron J. Komarek   Research Laboratories, Eastman Chemical Company, Kingsport, Tennessee

---

Current affiliation: Tokyo Institute of Technology, Midori-ku, Yokohama, Japan.

Marek Kowalczuk   Institute of Polymer Chemistry, Polish Academy of Sciences, Zabrze, Poland

Piotr Kurcok   Institute of Polymer Chemistry, Polish Academy of Sciences, Zabrze, Poland

F. Lefebvre   Chimie des Polymères et Systèmes Organisés, Université Libre de Bruxelles, Bruxelles, Belgium

Jacques Lemaire   Laboratoire de Photochimie, URA CNRS 433, Universite Blaise Pascal, Clermont Ferrand, and Laboratoire de Photochimie Moléculaire et Macromoléculaire, Ensemble Universitaire Cézeaux, Aubiere, France

Robert W. Lenz   Polymer Science and Engineering Department, University of Massachusetts, Amherst, Massachusetts

Abderrazek Maaroufi   Laboratoire de Chimie Organique et Biologique, URA CNRS 485, Ensemble Universitaire des Cézeaux, Aubiere, France

Minna Malin   Department of Chemical Engineering, Helsinki University of Technology, Espoo, Finland

Robert H. Marchessault   Department of Chemistry, McGill University, Montreal, Quebec, Canada

M. Masaoka   Central Laboratory, Rengo Co Ltd., Ohhiraki, Osaka, Japan

Jean M. Mayer   US Army Natick RD&E Center, Natick, Massachusetts

Stephen P. McCarthy   Department of Plastics Engineering, NSF Biodegradable Polymer Research Center, University of Massachusetts Lowell, Lowell, Massachusetts

Noriyuki Morohoshi   Tokyo University of Agriculture and Technology, Fuchu, Tokyo, Japan

R.-J. Müller   Biochemical Engineering, GBF, Gesellschaft für Biotechnologische Forschung mbH, Braunschweig, Germany

Kunio Nakamura   Otsuma Women's University, Chiyoda-ku, Tokyo, Japan

L. Nicolais   Department of Materials and Production Engineering and Institute of Composite Materials Technology, University of Naples "Federico II," Naples, Italy

S. Owen   Central Laboratory, Rengo Co Ltd., Ohhiraki, Osaka, Japan

Martin G. Peter[*] Institut für Organische Chemie und Biochemie der Universität, Bonn, Germany

C. A. Pettigrew The Procter & Gamble Company, Cincinnati, Ohio

Kaisa Poutanen VTT Biotechnology and Food Research, VTT, Finland

Sheldon D. Pratt University of Rhode Island, Narragansett, Rhode Island

Sathish Puthigae Department of Biochemistry, The Arrhenius Laboratories, Stockholm University, Stockholm, Sweden

G. A. Rece The Procter & Gamble Company, Cincinnati, Ohio

G. Romano Novamont, Novara, Italy

A. Rüdiger Department of Technical Microbiology, Institute of Biotechnology, Technical University Hamburg-Harburg, Hamburg, Germany

N. Sakota Central Laboratory, Rengo Co Ltd., Ohhiraki, Osaka, Japan

Mariastella Scandola Dipartimento di Chimica 'G. Ciamician' dell'Università di Bologna and Centro di Studio per la Fisica delle Macromolecole del C.N.R., Bologna, Italy

Thomas M. Scherer Department of Polymer Science and Engineering, University of Massachusetts, Amherst, Massachusetts

G. Schwarch University Montpellier 1, CRBA-URA CNRS 1465, Faculty of Pharmacy, Montpellier, France

Gerald Scott Polymer Processing and Performance Group, Department of Chemical Engineering and Applied Chemistry, Aston University, Birmingham, England

J. A. Scott Department of Chemistry, University of Toronto, Toronto, Ontario, Canada

Jukka V. Seppälä Department of Chemical Engineering, Helsinki University of Technology, Espoo, Finland

M. C. Smith The Proctor & Gamble Company, Cincinnati, Ohio

Armin Stein Institute of Organic and Macromolecular Chemistry, University of Jena, Jena, Germany

---

[*]*Current affiliation*: Institut für Organische Chemie und Strukturanalytik, Universität Potsdam, Potsdam, Germany.

Alexander Steinbüchel   Institut für Mikrobiologie der Westfälischen Wilhelms-Universität Münster, Munich, Germany

Barbara K. Sullivan   University of Rhode Island, Narragansett, Rhode Island

Chuanxin Sun   Department of Biochemistry, The Arrhenius Laboratories, Stockholm University, Stockholm, Sweden

A. Sunna   Department of Technical Microbiology, Institute of Biotechnology, Technical University Hamburg-Harburg, Hamburg, Germany

Graham Swift   Rohm and Haas Company, Spring House, Pennsylvania

Mark R. Timmins   Department of Polymer Science and Engineering, University of Massachusetts, Amherst, Massachusetts

Gianpaolo Tomasi   Dipartimento di Chimica 'G. Ciamician' dell'Università di Bologna and Centro di Studio per la Fisica delle Macromolecole del C.N.R., Bologna, Italy

M. Vert   University Montpellier 1, CRBA-URA CNRS 1465, Faculty of Pharmacy, Montpellier, France

Minna Vikman   VTT Biotechnology and Food Research, VTT, Finland

Alan W. White   Research Laboratories, Eastman Chemical Company, Kingsport, Tennessee

U. Witt   Biochemical Engineering, GBF, Gesellschaft für Biotechnologische Forschung mbH, Braunschweig, Germany

DEGRADABLE POLYMERS, RECYCLING, AND PLASTICS WASTE MANAGEMENT

POLYMER WASTE MANAGEMENT—BIODEGRADATION, INCINERATION, AND RECYCLING

SAMUEL J. HUANG

Polymer Science Program
Institute of Materials Science
University of Connecticut
Storrs, Connecticut 06269-3136

ABSTRACT

Increasing volumes of synthetic polymers are manufactured for various applications. The disposal of the used materials is becoming a serious problem. Unlike natural polymers, most synthetic macromolecules cannot be assimilated by microorganisms. Although polymers represent slightly over 10% of total municipal waste, the problem of non-biodegradability is highlighted by overflowing landfills, polluted marine waters, and unsightly litter. Existing government regulations in Europe and anticipated regulations in the United States will greatly limit the use of polymers in large volume applications (packaging, water treatment, paper and textile sizing, etc.) unless acceptable means of waste management are available. Total management of polymer wastes requires complementary combinations of biodegradation, incineration, and recycling. Biodegradation is the most desirable long-term future solution and requires intensive research and development before it becomes practical. On the other hand, incineration and recycling can become operational in a relatively short time for the improvement of the situation at present and in the near future.

Solid waste management is becoming increasingly difficult as traditional landfills are becoming scarce and, more importantly, environmentally undesirable. Among the problems is the ever-increasing volume of polymer wastes [1–4]. Most of the solid polymers (plastics) are used as protective coatings, structures, and

1

packagings. They are designed and manufactured to resist environmental degradations, including biodegradations. Since plastics are more economical than metals, woods, and glasses in terms of manufacturing costs, weight-to-strength ratio, and the amounts of energy and water required, as well as in most cases causing less environmental harm, the use of plastics is likely to increase. This makes polymer waste management an urgent problem, needing environmentally compatible and friendly solutions, both short and long term, as soon as possible [1–3].

In addition to conservation, direct waste management means are needed. Recycling, incineration, and biodegradation are possibilities. Among the advanced countries, Japan has made the most progress by far, both in terms of national policies and practice. Incineration has been and will continue to be the major means of waste management (Table 1). In Europe, Germany is actively pursuing mechanical recycling. In the United States, due to the lack of a national policy, there is hardly any practice of polymer waste management other than landfill, which is quickly becoming impractical.

In Japan, currently 11% of polymer wastes are recycled mechanically. This is limited mostly to industrial scrap. Consumer waste plastics recycling has been tried, but mostly abandoned. There do not seem to be any future plans in this area. Although around 65% of municipal solid wastes (MSW) in Japan is incinerated, only 15% is coupled with power generation. It is MITI's plan to increase plastic waste-to-energy conversation to 70% by the end of the 21st century and thus reduce the need of landfills to less than 10%. Recycling of industrial polymer scraps is being practiced, to a very limited extent, in Europe and the United States. This can be expected to increase as legislative incentives and economic benefits become more favorable. MSW consumer wastes contain polyethylene, polypropylene, polystyrene, and poly(vinyl chloride), which are incompatible with each other. As a result, only low performance and low market value products such as garden tires, fences, and planters can be manufactured. Compatibilization of these polyolefins is needed in order to make MSW plastics recycling practical.

Chemical recycling is potentially useful for certain polymers. At the present time, only poly(ethylene terephthalate) has been practically recycled [5]. Pyrolysis is another method by which small molecules can be obtained from polymer waste. New catalysts have to be developed for more efficient processes [6]. Little attention seems to have been directed toward catalytic pyrolysis of polyolefins, which should be similar to petroleum cracking processes.

Incineration is a common form of general waste disposal. Japan expects to eventually take care of up to 70% of its polymer wastes through incineration for energy processes [7]. In the United States, Connecticut is the only state in which 40–60% of its solid MSW is incinerated. However, because the MSW are not sorted,

TABLE 1.   Plastics Waste Management in Japan

	Current, %	21st Century (MITI), %
Material recycling	11	20
Thermal recycling	15	70
Landfill	37	< 10

40% of it ends up as ashes after incineration, and is difficult to dispose of [8]. Sorting polymer wastes before incineration will become necessary in the future. Since the combustion of relatively pure polyolefins generates too much heat for traditional furnaces, newer high temperature furnaces with ceramic liners might be needed. This will add significantly higher costs to the already very large costs for building incineration-to-energy plants. Gathering, sorting, and transportation are additional to the plant costs for incineration.

These facts, together with the growing concern over the greenhouse effect generated by the large volume of CO_2 produced, have caused almost insurmountable political barriers for building new incinerators in the United States.

Various forms of degradation can be used for polymer waste disposal. The most environmentally compatible is biodegradation, a subject of increasing research interest [9-19].

Most of the current large volume polymers are not biodegradable. Thus biodegradation for waste disposal can only become a reality when new biodegradable polymers and facilities for biodegradation become available. If biodegradations can be controlled and useful products can be obtained, they become bioconversion or

TABLE 2.

Pro	Con
Recycling (material, mechanical)	
Available processes	Product downgrade
Source reduction	Not easily adopted for mixed plastics
Suitable industrial scraps	High costs of gathering and sorting
Not final disposal	Not efficient for food packaging
Politically favored	Not a final disposal
Can be done on any scale	
Incineration	
High efficiency for sterilization	High plant cost
Energy generation	High gathering and sorting costs
Semifinal disposal	Could produce high water and gas pollution
Available technology	Political barrier
	Only applicable to relatively large scale
Biodegradation–Bioconversion	
Environmentally compatible and friendly	No enough reactors/plants
Completes the carbon and nitrogen cycles	Requires new plastics, additives, etc.
Can be on any scale	Has to overcome the public's misconceptions
	Needs to develop new products

biorecycling processes. Among the promising approaches to new useful biodegradable polymers are biopolymers, modified biopolymers, and blends.

Poly-R-3-hydroxyalkanoates are energy storage materials for bacteria and have been the subject of increasing research interest and limited commercial production [20–28]. Their biodegradation is being studied in detail [29–33]. If the processing of these polymers can be mastered, and if the costs of production can be lowered, they can become important biodegradable polymers.

Cellulose acetates are used for various applications. Recent results on their biodegradation [34, 35] have increased their importance as biodegradable materials. Starch derivatives, on the other hand, have not received as much attention.

Among the synthetic polyesters, polycaprolactone lacks the necessary high temperature properties, and polylactate cannot yet be produced cheaply. It is premature for these to become major polymers.

Blends of starch and degradable polymers are now commercially available [36–41]. Disposal facilities are needed for these to be commonly used.

SUMMARY

There are pros and cons for all three approaches (Table 2). Ideally, complementary practices of all three and conservation would be the most environmentally friendly in the long run.

REFERENCES

[1]   S. J. Huang, *Polym. Mater. Sci. Eng., 63*, 633 (1990).
[2]   J. Tallman, *Waste Age, 18*, 141 (1987).
[3]   B. Wessling, *Kunstoffe, 83*, 7 (1993).
[4]   A. M. Thayer, *Chem. Eng. News*, p. 7 (January 30, 1989).
[5]   R. Calendiene, M. Palmer, and P. von Bramiers, *Mod. Plast.*, p. 64 (1980).
[6]   J. A. Fiji, *Environ. Sci. Technol., 2*, 308 (1993).
[7]   *Plastic Wastes. Disposal and Recycling, Past, Present and Future in Japan*, Plastic Waste Management Institute, Tokyo, 1992.
[8]   K. Johnson, *New York Times*, p. B1 (January 30, 1989).
[9]   R. Leaversuch, *Mod. Plast. Int., 17*, 94 (1987).
[10] S. J. Huang, in *Encyclopedia of Polymer Science and Engineering*, Vol. 2, 2nd ed., Wiley, New York, 1985, pp. 220–243.
[11] S. J. Huang, in *Comprehensive Polymer Science*, Vol. 6 (G. Allen and J. C. Bevington, Eds.), Pergamon Press, London, 1989, pp. 567–607.
[12] D. L. Kaplan, J. M. Mayer, D. Ball, J. McCassie, A. L. Allen, and P. Sterhouse, in *Biodegradable Polymers and Packaging* (C. Ching, D. Kaplan, and E. Thomas, Eds.), Technomics Publishing, Lancaster-Basel, 1993, pp. 1–44.
[13] G. Swift, *Acc. Chem. Res., 26*, 105 (1993).
[14] R. Lenz, *Adv. Polym. Sci., 107*, 1–40 (1993).
[15] A.-C. Albertsson and S. Karlsson, in *Comprehensive Polymer Science*, First Supplement (G. A. Allen, S. L. Agarwal, and S. Russo, Eds.), Pergamon Press, London, 1993, p. 285.

[16] J.-C. Huang, A. S. Shetty, and M.-S. Wang, *Adv. Polym. Technol.*, *10*, 23 (1990).
[17] T. F. Cooke, *J. Polym. Eng.*, *9*, 171 (1990).
[18] P. J. Hocking, *J. Macromol. Sci. – Rev. Macromol. Chem. Phys.*, *C32*, 35 (1992).
[19] J. E. Potts, R. A. Clendinning, W. B. Ackart, and W. D. Niegisch, in *Polymers and Ecological Problems* (J. Guillet, Ed.), Plenum Press, New York, 1973, p. 61.
[20] E. A. Dawes and P. J. Senior, *Adv. Microb. Physiol.*, *10*, 135 (1993).
[21] Y. Doi, *Microbial Polyesters*, VCH, New York, 1990.
[22] A. J. Anderson and E. A. Dawes, *Microbiol. Rev.*, *54*, 450–472 (1990).
[23] P. A. Homes, S. H. Collins, and L. F. Wright, US Patent 4,477,654 (1984).
[24] A. Steinbüchel, *Biomaterials: Novel Materials from Biological Sources*, Stockton Press, New York, 1991, pp. 123–214.
[25] H. Brandly, R. A. Gross, W. R. Lenz, and R. C. Fuller, *Adv. Biochem. Eng./Biotech.*, *41*, 77 (1990).
[26] R. H. Marchessault, T. L. Blulim, Y. Deslandes, G. K. Hamer, W. J. Orts, P. R. Gundarajan, M. G. Taylor, S. Bloembergen, and D. A. Holden, *Makromol. Chem., Macromol. Symp.*, *19*, 235 (1988).
[27] G. J. M. de Konig, in *Unconventional and Nonfood Uses of Agricultural Biopolymers* (M. L. Fishman, R. B. Friedmand, and S. J. Huang, Eds.), American Chemical Society, Washington, D.C., 1994, In Press.
[28] M. K. Cox, in *Abstracts, 3rd International Scientific Workshop on Biodegradable Plastics and Polymers, November 1993, Osaka*, Japan Biodegradable Plastics Society, p. 32.
[29] Y. Doi, Y. Kanesawa, M. Kunioka, and T. Saito, *Macromolecules, 23*, 26 (1990).
[30] Y. Kawaguchi and Y. Doi, *Ibid.*, *25*, 2324 (1992).
[31] D. Jendrossek, I. Knoke, R. B. Habibian, A. Steinbüchel, and H. G. Schlegel, *J. Environ. Polym. Deg.*, *1*(1), 53 (1993).
[32] H. Nishida and Y. Tokiwa, *J. Environ. Polym. Degrad.*, *1*(1), 65 (1993).
[33] N. Nishida and Y. Tokiwa, *Ibid.*, *1*(3), 235 (1993).
[34] C. M. Buchanan, R. M. Gardner, and R. J. Komarck, *J. Appl. Polym. Sci.*, *47*, 1709 (1993).
[35] J.-D. Gu, D. T. Eberiel, S. P. McCarthy, and R. A. Gross, *J. Environ. Polym. Degrad.*, *1*(2), 143 (1993).
[36] C. Bastioli, V. Bellotti, L. Del Giudice, and G. Gilli, in *Biodegradable Polymers and Plastics* (M. Vert, J. Feijen, A. Albertsson, G. Scott, and E. Chiellini, Eds.), Royal Society of Chemistry, 1992, pp. 101–111.
[37] C. Bastioli, V. Bellotti, L. Del Giudice, and G. Gilli, *J. Environ. Polym. Degrad.*, *1*(3), 181 (1993).
[38] Y. Yoshida and M. Tomori, in *Abstracts, 3rd International Scientific Workshop on Biodegradable Plastics and Polymers, November 1993, Osaka*, Japan Biodegradable Plastics Society, p. 72.
[39] M. Koyama, H. Kameyama, and Y. Tokiwa, *Ibid.*, p. 71.
[40] S. Takagi, M. Koyama, H. Kameyama, and Y. Tokiwa, *Ibid.*, 71.
[41] *Michigan Biotechnology Institute Product Informations*, 1993.

DEGRADATION PRODUCTS IN DEGRADABLE POLYMERS

SIGBRITT KARLSSON and ANN-CHRISTINE ALBERTSSON

Department of Polymer Technology
Royal Institute of Technology
S-100 44 Stockholm, Sweden

ABSTRACT

Diffusion and migration of small molecules in polymers affect the long-term properties of the materials and also the surrounding environments. The interaction of such small molecules (i.e., low molecular weight compounds) with the environment are one of the main factors governing in-vitro and in-vivo behavior of degradable polymers. The type and amount of formed degradation products are the crucial points deciding the applicability of degradable polymers. The number of low molecular weight compounds obtained during long-term use of degradable polymers depends on the polymeric chain. Hydrolyzable polymers like the polyesters give few degradation products (mainly the monomer) where it is possible to relate the amount of products formed with the weight loss and the molecular weight changes. On the other hand, nonhydrolyzable polymers like polyethylene form hundreds of products in different amounts, thus complicating this comparison. Degradation products of hydrolyzable polyesters (PHB, PLLA, PDLLA/PGA, and PAA) as obtained by (head-space)-GC-MS are presented. The concept of fingerprinting based on the abiotic and biotic degradation products formed in degradable LDPE (LDPE + starch + prooxidant) is described and can be related to our proposed biodegradation mechanism of polyethylene.

7

INTRODUCTION

Increased interest in the use of degradable polymers has led to the need for accurate and precise test methods. Parallel with the demand for controllable and quick degradation is the demand for the development of nonharmful low molecular weight compounds which can be part of natural cycles. In this context the questions to be addressed are the type and amount of single products and the total amount of low molecular weight compounds present in degradable polymers.

The interaction of low molecular weight compounds from degradable polymers is one of the major factors governing the in-vitro and in-vivo behavior of materials. There are several types of small molecules in polymers. Some are present from the beginning, i.e., synthesis related (trace monomers, catalysts, solvents, and additives). Other products can be related to the type of processing and the processing conditions. Yet others are formed during use; they are usually referred to as degradation products. These degradation products are complex mixtures of compounds formed during biodegradation, thermooxidation, hydrolysis, etc. Degradation products of additives are also generally identified. Different polymers form different numbers of degradation products, ranging from the monomer and oligomer of the monomer (e.g., polyesters) up to hundreds of products (e.g., polyethylene).

Diffusion, migration, and leakage of additives and/or degradation products increase the degradation rate and also leave the polymeric matrix more brittle, thus making the material more prone to further degradation. The monitoring of small molecules in recycling is important as these will affect the new products and their performance. Continuous loss of additives demands addition of new additives in order to fulfill the material properties.

In this paper some results are presented concerning the detection and identification of degradation products in degradable polymers. Examples will be given from polyesters [polyhydroxyalkanoates (PHA), poly(lactic acid) (PLA), PLA copolymerized with poly(glycolic acid) (PGA) and poly(adipic anhydride) (PAA)]. The concept of fingerprinting is discussed for the first time in connection with synthetic degradable polymers to describe abiotic and biotic degradation products of LDPE + starch + prooxidant (i.e., LDPE + 20% MB) and related to our proposed biodegradation mechanism of polyethylene.

EXPERIMENTAL

Samples and Degradations

Commercial Biopol MBL granules of PHB and PHB-*co*-PHV (7% PHV) were dissolved in chloroform at 20 g/L by refluxing. To this solution was added nine times its volume of methanol, and the precipitated polymer was filtered and washed with methanol. The purified granules were thermooxidized. PHB (200 mg) was aged in sealed head-space vials at 100°C for up to 500 hours.

PLLA (100 mg) and PDLLA/PGA (75/25) were immersed in phosphate buffer (described elsewhere [1]) of pH 7.4, incubated at three different temperatures (37, 60, and 90°C), and aged for periods up to 60 days. The PAA samples were polymerized from oxepan-2,7-dione in our laboratory and immersed in 10 mg of phosphate buffer as described above. Samples were withdrawn at regular intervals for periods up to 35 hours.

LDPE films (80 μm) were made by a conventional blown film process using a Betol extruder with a 25-mm screw of L:D 20:1, a blow-up ratio of about 2.5:1, and a die temperature of 185°C [2]. The pro-degradant additives were incorporated into the LDPE matrix in the form of a masterbatch in the amount of 20%, mostly consisting of corn starch (7.7%), styrene–butadiene copolymer (SBS), maganese stearate, and LLDPE. The degradation took place according to the following procedure. The films were preoxidized by heating in air at 100°C for 6 days in order to surpass the induction period and ensure that oxidation of the LDPE matrix had commenced. Thereafter the films were subjected to either sterile aqueous media [2] or to the bacteria *Arthrobacter paraffineus* (biotic media). Incubation took place at ambient temperature for periods up to 15 months.

Extractions

Acidified buffer solutions of PLLA, PDLLA/PGA, and PAA were extracted three times with diethylether (p.a.). The extracts were combined and evaporated by a stream of nitrogen. The same type of extraction was performed with the LDPE–20% MB samples [2].

Derivatization

PLLA, PDLLA/PGA, and PAA extracts needed to be derivatized before GC-MS analysis, and this was done as follows. 20 μL MTBSTFA (*N-tert*-butyldimethylsilyl)-*N*-methyltrifluoroacetamide, 98%, was added to the dry residue of the extracts, and the solutions were diluted with 0.1 mL isooctane. The reaction was complete after 30 minutes at 60°C. The derivatized samples were injected into the GC-MS.

Analysis

The head-space (Hs)-GC-MS analysis was carried out using a Perkin-Elmer 8500 GC to which a head-space unit HS 101 and an ion trap detector (MS) were connected. The column was a Chromopack wall-coated open tubular fused silica, 25 m × 0.32 mm (CP-SIL-19 CB), and a temperature program from 40 to 180°C at 5°C/minute were used. The HS unit was thermostated for a period of 50 minutes at 100°C. The ITD operating conditions were: mass range, 20–600 u; scan time, 1.000 second; peak threshold, 1 count; mass defects, 0.03 mmu/100 amu; multiplier voltage, 1600 V; and a transfer line temperature of 250°C. The temperature program for PLLA, PDLLA/PGA, and PAA was an oven temperature 40°C (1 minute), programmed to 200°C at 10°C/min, and then isothermic temperature of 200°C (8 minutes). The derivatized samples were separated on a CP-SIL-43 CB column.

The LDPE–20% MB extracts were injected into a Varian 3400 GC equipped with a J&W 30 m × 0.32 mm DB-5 column, film thickness 0.25 μm, and a flame ionization detector (FID). Oven temperature was held at 50°C at 1 minute, raised at a rate of 5°C/min to 310°C, and then held for 10 minutes. Nitrogen was used as the carrier gas. Identification was made by comparison with retention indices from standard compounds.

The SEC analysis was performed using a Waters Associates M-6000A size-exclusion chromatograph equipped with a set of Styragel columns, thermostated with a 1 mL/min flow, with chloroform as the mobile phase. The molecular masses [number average (M_n) and mass average (M_w)] were calculated using a calibration curve for polystyrene.

RESULTS AND DISCUSSION

The polyhydroxyalkanoates (PHA) represent a large group of polyesters produced from renewable resources. This fact, together with their degradability, makes them interesting materials for packages, etc. The degradation products from poly(β-hydroxybutyrate) (PHB) are fairly simple, and the monomer and oligomers of the polymer can be identified [3]. The conversion of β-hydroxybutyrate to crotonic acid is so fast that often only crotonic acid can be identified [3]. The hydrolysis of PHA can be done in different media and at different temperature, and the amount of products formed is dependent on the pH, the temperature, and the duration of degradation [3]. The increased degradation rate at higher pH depends partly on the increasing solubility of the polymeric degradation products at higher pH in addition to the overall increased hydrolysis rate.

Figure 1 shows the production of crotonic acid as a function of thermooxidation time at 100°C as obtained by head-space GC-MS. The same figure also shows the changes in molecular weights (M_n) during degradation. The lowest value of M_n is noted for those samples where a large amount of crotonic acid was formed. In addition 2-ethoxyethyl acetate and acetaldehyde were identified in thermooxidized native polyesters. The thermal decomposition of PHB and its copolymers is the result of random chain scission occurring by a six-membered ring ester decomposition process to form carboxylic ends in the polymer.

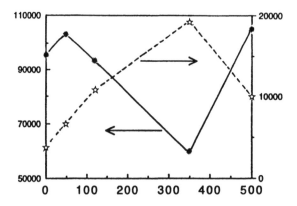

FIG. 1. M_n (✳, solid line) and amount of crotonic acid (✩, broken line) produced as a function of aging time of PHB at 100°C. y-axes in surface units (to the right) and g/mol (to the left).

Figure 2 shows the formation of lactic acid (LA) in aged PLLA (phosphate buffer) at three different temperatures. In the 37 and 60°C aged samples, a steady state in the formation of LA is quickly reached; the 90°C aged samples releases LA in a continuously increasing way. The weight loss of PLLA materials follows a similar pattern; 5 days of aging at 90°C corresponds to a 40% weight loss. A 75/25 copolymer of PDLLA and poly(glycolic acid) (PGC) releases about twice as much lactic acid during the same hydrolysis time as the PLLA homopolymer. This parallels the observations of the molecular weight reductions. It was observed by SEC that the molecular weight changes are largest during the first 10 days and that copolymers with glycolic acid give a material with increased susceptibility to hydrolysis at different temperatures. The onset of a rapid evolution of lactic acid is simultaneous with a drastic decrease in molecular weight.

Figure 3 presents the volatile degradation product of poly(adipic anhydride) (PAA) obtained by head-space GC-MS. Contrary to the polyesters presented in Figs. 1 and 2, this polymer degrades very quickly and has therefore found application in slow release formulations. Only one degradation product (adipic acid) was identified in amounts increasing with hydrolysis time. The formation of AA from hydrolyzing PAA is parallel with the detection of released pharmaceutical. The increase in the amount of evolved AA per milligram of PAA is quick; a steady-state level is reached after 10 hours.

In microbiology and bacteriology it is customary to identify the type of microorganism responsible for a certain disease by detecting its degradation products [4, 5]. Different microorganisms have different metabolisms (or degradation mechanisms), and this is reflected by their degradation products. This is referred to as

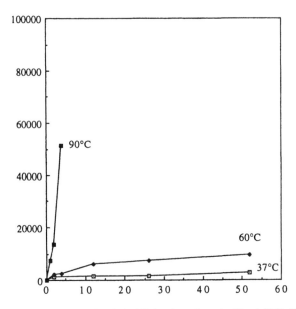

FIG. 2.   Hydrolysis of poly(lactic acid) in phosphate buffer (pH 7.4). The amount of lactic acid detected after different hydrolysis times and at different temperatures. Analysis performed by ion-trap GC-MS instrument. x-axis = hydrolysis time (days); y-axis = absorption units (lactic acid).

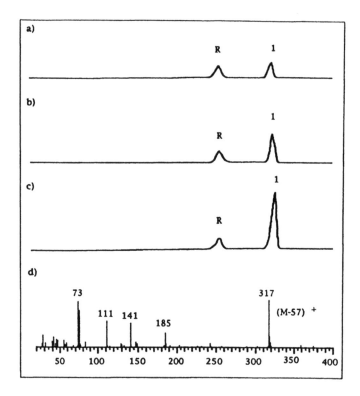

FIG. 3.   Head-space GC-MS chromatograms of hydrolyzed PAA. All samples were derivatized with MTBSTFA before analysis. Chromatograms of released adipic acid after a) 0.5 hour, b) 2 hours, and c) 6 hours in phosphate buffer at 37°C (pH 7.4). d) Mass spectra of TBDMS ester derivative of AA, Peak 1. R = Reagents.

"fingerprinting" and can also be applied to the degradation of synthetic polymers in different environments. Figure 4 displays chromatograms of the degradation products of LDPE + 20% MB aged for 10 weeks in aqueous biological and sterile environments. Figure 4(a) represents the fingerprint of a biologically degraded sample. The absence of a substantial amount of carboxylic acid should be noted; instead, different late esters are found. The sterile sample contains several peaks identified as carboxylic acid (Fig. 4b). Both chromatograms agree well with the LDPE biodegradation mechanism proposed by us in 1987 [6].

In parallel with the use of a new, specialized chromatography, we have developed a sophisticated prepreparation process to avoid the loss of different classes of compounds. It is based on solid phase extraction which is superior to ordinary liquid extraction. We have succeeded in identifying over 50 different degradation products in aged LDPE + MB samples [2]. Our research has also demonstrated that comparison with molecular weight changes, weight losses, and/or FT-IR is fruitful in properly describing the degradation mechanisms. The present results also show that the product pattern can be used as fingerprints, e.g., look for the presence or absence of special groups of homologous series (volatile acids, hydrocarbons, etc.). Thus, the fingerprints correlate with the degradation mechanisms.

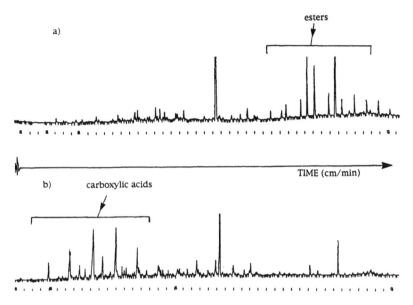

FIG. 4.   GC chromatograms of degradation products evolved in aqueous biotic environments and aqueous abiotic environment from LDPE + 20% masterbatch (starch + prooxidant). a) biotic environment and b) abiotic environment. The x-axis is the time scale.

CONCLUSIONS

Specialized chromatography (head-space GC-MS, LC, etc.) has been developed allowing the detection and identification of low molecular weight compounds in degradable polymers. We have demonstrated that the product pattern can be used as fingerprints and that the product pattern is related to the degradation mechanism and the kinetics. In addition it was also shown that the type of compounds formed is highly dependent on the environment.

The interaction of polymers with the environment governing the in vitro and in vivo behavior of degradable polymers is partly due to the evolution of low molecular weight compounds. The crucial point when using degradable polymers is, thus, the determination of the interaction of degradable polymers with the environment as manifested by the evolution of degradation products.

REFERENCES

[1]   A.-C. Albertsson and O. Ljungquist, *Acta Polym., 39*, 95 (1988).
[2]   A.-C. Albertsson, C. Barenstedt, and S. Karlsson, *Ibid., 45*, 97 (1994).
[3]   S. Karlsson, C. Sares, R. Renstad, and A.-C. Albertsson, *J. Chromatogr., 669*, 97 (1994).
[4]   S. Karlsson, Z. G. Banhidi, and A.-C. Albertsson, *Ibid., 442*, 267 (1988).
[5]   S. Karlsson, Z. G. Banhidi, and A.-C. Albertsson, *Appl. Microbiol. Biotechnol., 28*, 305 (1988).
[6]   A.-C. Albertsson, S. O. Andersson, and S. Karlsson, *Polym. Degrad. Stab., 18*, 73 (1987).

RECYCLING BIOPOL—COMPOSTING AND MATERIAL RECYCLING

M. K. COX

Biopolymers Group
ZENECA Bio Products
PO Box 2, Belasis Avenue, Billingham, Cleveland TS23 1YN, England

ABSTRACT

The recycling of biodegradable thermoplastics such as ZENECA's BIOPOL range of poly-3-hydroxybutyrate and poly-3-hydroxyvalerate copolymers needs to be considered in terms of both material recycling and organic recycling by composting. BIOPOL can be recycled as regrind. The addition of BIOPOL to a model waste stream demonstrates that at the anticipated addition levels, BIOPOL should not have a deleterious effect on the processing or properties of the thermoplastic waste stream. Therefore, BIOPOL should not compromise the recycling initiatives underway. Recycling BIOPOL by composting not only offers an additional waste management option where material recycling is not a viable option for technical, economic, or environmental reasons, but also generates a useful product, valuable as a soil conditioner.

INTRODUCTION

There has been an increase in both the quantity of waste generated and public concern over its effect on the environment. Plastic packaging waste is a particularly visible problem and represents 20–30% of the waste stream by volume (although only ~7% by weight). It is generally recognized that there is no single, simple solution to waste management. A range of waste management options exist which either reduce the amount of material entering the waste stream or deal with its

15

disposal. A hierarchy of waste management options appears to be arising which consists of reduce, reuse, recycle (including composting), incineration, and landfill, in decreasing order of desirability. These options are applicable to plastic waste.

Up to 60% of municipal solid waste is organic and potentially compostable. Composting has significant potential to not only offer an alternative waste disposal option for a significant proportion of waste, but also to generate a useful product, valuable as a soil conditioner. Composting can be considered to be an organic recycling process. The use of biodegradable polymers in packaging organic/biodegradable contents such as food potentially enables all of the resultant waste to be composted without requiring separation of contents and packaging. Composting is particularly pertinent where material recycling is not a viable option for technical, economic, or environmental reasons. A significant increase in the levels of composting is expected to result in a significant diversion of material from landfilling.

ZENECA's BIOPOL range of poly-3-hydroxybutyrate and poly-3-hydroxyvalerate bacterial copolyesters are compatible with the range of waste management options. It has been demonstrated that BIOPOL can be recycled in the traditional sense as regrind or in a model mixed plastic waste stream without adversely affecting the processing of properties of the latter and therefore should not compromise the recycling initiatives underway. In addition, BIOPOL can be organically recycled by composting. Indeed, both forms of recycling can be practiced. BIOPOL can be initially materially recycled (say as regrind) and then finally organically recycled by composting. Therefore, the recycling of BIOPOL should be considered in terms of both material recycling (as for conventional thermoplastics) and organic recycling by composting (which requires biodegradation).

MATERIAL RECYCLING

Material recycling involves the reprocessing of the polymer, either alone or with other polymers.

During melt processing of a polymer by, say, injection molding, waste is produced in the form of sprues and runners and substandard parts. This waste can be readily collected and kept free of contamination, and is, therefore, of good quality provided excessive thermal degradation has not occurred during processing. The sprues, etc. are reground, mixed with virgin material, and melt processed into finished articles. This use of regrind is a common form of material recycling to reduce the material costs of production. Typically 20% regrind levels are used.

Once the articles have been sold, used, and disposed of, they can then also potentially be recycled. To recycle the polymer alone it must be collected, identified, and separated from the other waste. Excessive contamination of the polymer (e.g., with food) and the use of composite articles may prevent cost-effective recycling of this postconsumer waste.

Recycling of a mixed plastic waste stream avoids the need to identify a specific plastic and separate different plastics, but tends to produce a lower quality, variable product. Mixed plastic waste streams therefore tend to be used for relatively undemanding applications.

BIOPOL Regrind

BIOPOL is a thermoplastic, and regrind can be added and mixed with virgin material, as for conventional thermoplastics. Regrind levels of up to 20% are generally recommended for BIOPOL.

Experimental evidence has been obtained to demonstrate that BIOPOL regrind can be used.

Unfilled BIOPOL

BIOPOL copolymer with 11% hydroxyvalerate (HV) content was subjected to a number of processing (melt extrusion/granulation) cycles. After each extrusion/granulation cycle a sample was injection-molded and mechanical properties measured. This therefore represents the use of 100% BIOPOL regrind at each stage. Up to four reprocessing cycles were used; that is, reprocessing 100% BIOPOL regrind up to four times. This is therefore a severe test of the suitability of BIOPOL regrind for general use in melt processing BIOPOL.

The data show that reprocessing 100% BIOPOL regrind four times had little deleterious effect on tensile strength, elongation at break, flexural modulus, flexural strength, and heat distortion temperature. As expected, some decrease in impact strength was observed, with four reprocessing cycles of 100% BIOPOL regrind reducing notched Izod impact strength by ~ 18%.

BIOPOL Bottle Grades

Bottles were extrusion blow-molded using virgin and 100% regrind in BIOPOL grades D410G and D411G. Horizontal and vertical test pieces were then cut out of the bottles and mechanical properties measured. Again, this is a severe test of the use of BIOPOL regrind since levels of up to 20% are recommended, and 100% regrind was used in these experiments.

The data for the virgin and 100% regrind bottles show that values for tensile strength, elongation at break, and modulus are similar. As expected, some decrease in energy to break of approximately 20% was observed. Scatter in the data was evident, which was expected since test pieces were cut from actual bottles.

BIOPOL in a Mixed Plastic Waste Stream

The wide range of properties required from polymers for packaging and other applications necessitates the use of a wide range of polymers. These eventually arise in the waste stream. Since the properties of a mixed plastic waste stream are variable and somewhat inferior, with consequent limited applications, the emerging recycling industry aims to recycle the major bulk commodity plastics as separate waste streams. The Society of Plastics Industry (SPI) Classification divides the recyclable thermoplastic waste stream into seven broad categories. Class 1 is polyethylene terephthalate (PET), Class 2 is high density polyethylene (HDPE), Class 3 is polyvinyl chloride (PVC), Class 4 is low density polyethylene (LDPE), Class 5 is polypropylene (PP), Class 6 is polystyrene (PS), and Class 7 contains all other thermoplastics. Currently, BIOPOL is included in Class 7.

Sales estimates for the year 2000 of conventional plastics and BIOPOL indicate that BIOPOL will represent much less than 1% of the total thermoplastics waste stream. In practice, BIOPOL would be expected to be recycled with Class 7 materials, for which by the year 2000 it would represent less than 1% of the total. At these low levels, any effect of BIOPOL on the quality and properties of such mixed waste streams is expected to be minimal.

This was tested experimentally by preparation of a model mixed polymer waste stream. 1% BIOPOL copolymer (with 10% HV content) was added to a mixed polymer blend containing 40% LDPE, 25% PP, 15% high impact PS, 10% PVC, 2.5% acrylonitrile-butadiene-styrene (ABS), 2.5% nylon, 2.5% polycarbonate (PC), and 2.5% polyacetal. The resultant blend was extruded, granulated, and injection-molded at temperatures up to 245°C with no significant difficulties. The model mixed polymer waste without BIOPOL exhibited an impact strength of 39.3 J/m, a tensile strength of 14.1 MPa, and a Young's modulus of 393 MPa compared with values of 39.8 J/m, 13.2 MPa, and 397 MPa, respectively, for the waste plus 1% BIOPOL. This confirms that the effect of BIOPOL on a mixed polymer waste stream is minimal at levels of 1%.

On rare occasions it may be possible for relatively higher levels of BIOPOL to be present in a mixed polymer waste stream. A large range of blends of BIOPOL and other thermoplastics have been investigated within ZENECA and externally which demonstrate the effect of varying BIOPOL levels on blend properties. These experiments have also demonstrated that BIOPOL can be successfully melt-processed over a range of temperatures with a wide range of polymers, including LDPE, PP, PS, PVC, ABS, nylon, polyacetal, PC, polyvinyl alcohol, polycaprolactone, and high-density polyethylene (HDPE). BIOPOL is melt processible in conventional extruders and does not require special preparation prior to melt processing.

BIOPOL bottle grade D411G white 1001 and unfilled HDPE were tumble blended to produce blends with 1, 5, and 10% by weight BIOPOL, extruded and injection-molded under the same conditions as a HDPE control. BIOPOL was readily melt processed with HDPE at levels up to 10% (the maximum used in this experiment). The addition of 1% BIOPOL to HDPE had little effect on melt flow index (MFI) and mechanical properties (tensile strength, Young's modulus, and Izod impact strength). Addition of 5% and above of BIOPOL to HDPE caused some loss of mechanical properties (particularly impact strength), but useful properties were still maintained at the 10% level.

Higher levels of BIOPOL were readily melt-blended with LDPE. Addition of 20, 30, and 40% BIOPOL to LDPE progressively increased modulus and decreased tensile strength and elongation at break.

Addition of BIOPOL to certain polymers such as PVC and polyacrylonitrile has a beneficial effect.

BIOPOL appears to be at least partially miscible with PVC, and blends have been successfully extruded and molded. BIOPOL has shown some promise as a processing aid in rigid PVC formulations. Addition of 20% BIOPOL to PVC (by weight) increased the falling weight impact energy to break for 3.9 to 24.0 J.

BIOPOL also appears to be effective as a processing aid for polyacrylonitrile, with addition of small quantities significantly reducing the power required for extrusion. Addition of 2 phr (parts per hundred of resin) of BIOPOL to a 35% nitrile group containing polyacrylonitrile increased Young's modulus, tensile strength, elongation to break, and notched Izod impact strength (from 3.63 GPa, 57 MPa, 10-20%, and 110-150 J/m to 4.18 GPa, 68 MPa, 33-36%, and 370-500 J/m, respectively).

Biodegradation studies on blends of BIOPOL with other polymers have been undertaken within ZENECA and externally. Addition of BIOPOL to a nonbiodegradable polymer such as PP does not cause biodegradation of the PP. In soil burial tests at 25°C, the weight loss after 63 days of 100% BIOPOL and 10% BIOPOL/ 90% PP blend films was 71.4 and 0.14%, respectively. This shows that when BIO-POL is present in small quantities (≤ 10%) and encapsulated within the nonbiodegradable PP matrix, the microorganisms cannot readily access the BIOPOL and biodegradation of the BIOPOL is rendered insignificant. The PP does not biodegrade. Therefore the presence of small quantities of BIOPOL in a mixed plastic waste stream should not affect stability with respect to biodegradation.

Blends of BIOPOL with other biodegradable polymers such as polycaprolactone biodegrade as expected.

In conclusion, BIOPOL is likely to be recycled in the SPI Class 7 category and at the anticipated level of < 1% is not expected to significantly adversely affect the melt processing of the mixed plastic waste stream or its properties (including stability to biodegradation). Naturally, as for other polymers, the inclusion of BIOPOL in a separate pure waste stream such as PET should be avoided.

COMPOSTING—ORGANIC RECYCLING

Composting may be defined as the aerobic biodegradation of organic material to form primarily carbon dioxide, water, and humus.

Like material recycling, organic recycling of biodegradable materials such as BIOPOL requires collection and separation (from nonbiodegradable impurities). The large (up to 60%) quantity of municipal waste that is organic and therefore potentially compostable is encouraging the continued development of composting facilities. Use of BIOPOL packaging for compostable waste simplifies the composting process by avoiding the need to separate the contents from the packaging. In addition, use of BIOPOL bags for waste further simplifies the process, since the entire bag and contents can be added to the compost stream.

The organic recycling of BIOPOL by composting has been extensively studied at the industrial scale and under a range of composting conditions.

Initial studies conducted at an industrial composting plant in Germany indicated that up to 80% of the weight of BIOPOL bottles was biodegraded or fully incorporated into compost within 15 weeks. Any remaining material consisted of small fragments with signs of active degradation on the surface.

Recently the composting of BIOPOL has been extensively studied over a wide range of conditions (which may be encountered in practice) by the Ingenieurgemeinschaft Witzenhausen (igw-Kompostverwertung)/Plan Co Tec with Kassel University in Germany. A full report will be issued. In summary, BIOPOL is compostable over a wide range of temperatures and moisture levels, with maximum biodegradation rates occurring at moisture levels of ~ 55% and temperatures of ~ 60°C. These conditions are similar to those observed in most large-scale composting plants. No significant differences were observed in the biodegradation rates between the different BIOPOL grades evaluated.

Up to 85% of the BIOPOL samples degraded within 7 weeks under these controlled composting conditions. In addition, BIOPOL-coated paper was rapidly degraded and assimilated completely into compost.

The quality of the final compost is judged by the growth of seedlings in it. Seedling growth in control compost was compared to that in mixtures of control compost and 25% BIOPOL compost. Plant growth of over 90% that observed in the control compost is considered to indicate compost of acceptable quality. Composts containing 25% BIOPOL compost produced seedling growth approximately 125% of that in the control compost. Therefore, not only is BIOPOL compostable, but the resultant compost is of high quality, supporting a relatively high level of seedling growth.

CONCLUSIONS

BIOPOL is compatible with a wide range of disposal and waste management options. BIOPOL can be recycled through two distinct routes, material recycling and organic recycling (composting).

BIOPOL regrind can be added to virgin material for material recycling, as with other plastics, with little effect on mechanical properties. Regrind levels of up to 20% are recommended.

Addition of the anticipated levels of BIOPOL to a mixed plastics waste stream should not have a deleterious effect on the properties or processing of the thermoplastic recycling waste stream.

Organic recycling by composting is applicable for up to 60% of municipal solid waste and can generate a useful product, valuable as a soil conditioner. Addition of BIOPOL to a composting stream produces compost of high quality which supports a relatively high level of seedling growth.

Organic recycling by composting offers an alternative waste management option where material recycling is not a viable option for technical, economic, or environmental reasons.

LABORATORY-SCALE COMPOSTING TEST METHODS TO DETERMINE POLYMER BIODEGRADABILITY: MODEL STUDIES ON CELLULOSE ACETATE

RICHARD A. GROSS and JI-DONG GU†

Department of Chemistry

DAVID EBERIEL

Department of Biology

STEPHEN P. McCARTHY

Department of Plastics Engineering

NSF Biodegradable Polymer Research Center
University of Massachusetts Lowell
One University Avenue, Lowell, Massachusetts 01854

ABSTRACT

Studies have been conducted to determine the time dependence of film normalized weight loss using in-laboratory simulations of a compost environment. The composting bioreactors contained a fresh synthetic waste mixture formulation which was maintained at 53°C and 60% moisture. An important result from these studies is that cellulose acetate (CA) films (approximate film thickness of between 0.025 and 0.051 mm) with degrees of substitution (DS) of 1.7 and 2.5 appeared completely disintegrated after 7 and 18 day exposure periods, respectively. Little weight loss of these films was noted under similar temperature and

†Present address: Laboratory of Microbial Ecology, Division of Applied Sciences, 40 Oxford Street, Harvard University, Cambridge, Massachusetts 02138.

21

moisture conditions using abiotic controls. The testing protocol which was developed provided repeatable results on normalized film weight loss measurements. The dependence of the synthetic waste formulation used on polymer film weight loss was investigated using, primarily, CA DS-1.7 as a substrate. It was found that the time dependence of CA DS-1.7 film weight loss was virtually unchanged for five of the seven formulations investigated. However, two formulations studied resulted in relatively slower CA DS-1.7 film weight loss. Decreasing moisture contents of the compost from 60 to 50 and 40% resulted in dramatic changes in polymer degradation such that CA DS-1.7 polymer films showed an increase of the time period for a complete disappearance from 6 to 16 and 30 days, respectively. Also, a respirometric test method is described which utilizes predigested compost as a matrix material to support biological activity. Using this method, conversions to greater than 70% of the theoretical recovered CO_2 for CA (1.7 and 2.5 DS) substrates were measured. Therefore, these results indicate high degrees of mineralization for CA with DS $\leq$ 2.5 under the appropriate disposal conditions.

INTRODUCTION

There is an increased recognition that biodegradable plastics can serve an important role in the design of an intelligent, integrated, solid waste disposal scheme [1–3]. The concept is elegant, achievable in principle, and environmentally appealing. Biodegradable disposable-plastic articles can be designed such that they will be entirely converted by microbial activity in a biologically active environment to biogas (CO_2 and CH_4/CO_2 under aerobic and anaerobic conditions, respectively), biomass, and biological by-products. This process, which is also called mineralization, should be considered as a biologically mediated recycling of plastic articles.

Composting of organic wastes has been recognized as an important technology that produces organic humus materials. This compost product can be marketed and used successfully for the reclamation of mined soil, highway construction, home use, and cover for landfills [4–7]. Important work has been conducted in obtaining an understanding of the microbiological, physical, as well as chemical changes in compost environments both in the laboratory [8–16] and from outdoor studies [6, 17–23]. The composting principle has expanded into pesticides and hazardous chemicals management [24, 25]. In addition, the CO_2 evolution resulting from the microbial degradation of a chemical organic compound is used by the US Food and Drug Administration for regulatory purposes [26]. Since composting of organic solid wastes has excellent potential for future growth as a component in the solid waste solution, it then becomes desirable to obtain rates both for the loss in physical integrity (termed biodestructability herein, often measured in weight loss) and mineralization of both disposed plastic articles and water-soluble polymers.

The study of polymer degradability using in-laboratory accelerated test methods which are designed to simulate actual disposal conditions is an approach currently under development by us and others. Laboratory-scale testing methods suitably address many concerns with the use of outdoor test exposures where only limited control is possible of the environmental conditions used during testing [27–30].

The question that must be answered when in-laboratory simulation test methods are used is to what degree of certainty do they predict polymer degradability in actual disposal sites. Furthermore, some investigators who have adopted this approach for plastic degradation testing use materials to create the waste matrix supporting biodegradability that are not well defined, difficult to obtain by other laboratories, and by their very nature highly variable [31, 32]. Work described herein begins to probe the importance of environmental parameters such as the use of synthetic waste mixtures of variable nutrient content as well as the effects caused by the compost percent moisture on polymer biodegradability. It is expected that through such work and the development of appropriate mathematical models it will eventually be possible to correlate results obtained from in-laboratory testing to expected rates of degradation of plastics in variable outdoor compost sites.

In this paper, laboratory-scale composting test methods to determine the weight loss of plastic articles as well as the percent conversion of polymer carbon to CO_2 are presented. In addition, studies were carried out to determine the influence of the compost mixture composition and percent moisture on the rate of film degradation. For the purpose of this paper, the results presented are limited to those obtained using films of cellophane and cellulose acetate (CA) with degrees of substitution (DS) of 1.7 and 2.5.

MATERIALS AND METHODS

Materials Used, Film Preparation, and Characterization

CA DS-1.7 pellets and DS-2.5 powder with no additives were provided by the Eastman Kodak Company (Kingsport, Tennessee). The polymers, as received, had weight-average molecular weights ($\overline{M}_w$) and dispersities ($\overline{M}_w$/number-average molecular weight, $\overline{M}_n$) of 200,000 and 350,000 g/mol, respectively, and 1.7 and 1.8, respectively. As was previously described [9, 11], the molecular weight and dispersity values were determined by GPC relative to polystyrene standards, and the degree of substitution specified by the manufacturer was confirmed by ^{1}H-NMR measurements. CA films were made by solution casting onto Teflon from either acetone–H_2O (1:1 ratio, v/v; 17% CA DS-1.7, w/v) or acetone (15% CA DS-2.5, w/v). The film thickness values for CA DS-1.7 and DS-2.5 were ~0.013 to ~0.025 and ~0.051 mm, respectively, unless otherwise specified. These films were aged for a minimum of approximately 2 weeks prior to use. Cellophane film, noncoated, with a thickness of approximately 0.041 mm, was obtained from E. I. du Pont de Nemours & Company (Wilmington, Delaware). The films were cut into dimensions of 20 × 20 mm, dried in a vacuum oven at 30°C (10 mmHg) to constant weight, the weight was recorded, and the films were exposed to the test environments.

Laboratory-Scale Aerobic Composting Using Fresh Waste Mixtures to Study Plastic Degradation

General Methodology

A simulated *in-laboratory* compost procedure has been developed and used to study the degradability of shaped polymeric materials [9–11]. In this procedure, synthetic municipal solid waste formulations were used (see below and Table 1) that

TABLE 1.   Contents of Synthetic MSW Formulations Used to Carry Out Laboratory-Scale Compost Degradation Testing

Name of components	Compost mix formulation[a]						
	Mix 1	Mix 2	A	B	C	D	E
Tree leaves[b]	45.0	47.0					
Shredded paper[c]	16.5	17.0	17.0	33.0	33.0	33.0	33.0
Food waste	6.7	7.0	7.0				
Meat waste	5.8	6.0	6.0				
Cow manure[d]	17.5	18.0	18.0	15.0	15.0		10.0
Saw dust	1.9	2.0	2.0				
Aluminum and steel shavings	2.4						
Glass beads	1.3						
Urea	1.9	2.0	2.0				3.0
Steam-exploded wood			47.0	33.0	33.0	33.0	33.0
Dried timothy				18.0			
Dried alfalfa					18.0		
Rabbit choice						33.0	
Starch							20.0
Compost inoculum	1.0	1.0	1.0	1.0	1.0	1.0	1.0
C/N ratio	14.0	13.9	21.2	61.4	38.2	49.9	24.7

[a]All numbers are weight percentage values.
[b]Shredded maple and oak leaves at 1:1 ratio.
[c]Shredded news and computer papers at 1:1 ratio.
[d]Dehydrated cow manure.

contain important components present in the municipal solid waste stream [12a]. An experimental study for a specific polymer or composting condition was carried out in triplicate in bioreactors along with a poisoned control test vessel. The control vessel was poisoned after autoclaving by the addition of KCN. Triplicate plastic samples were removed from each composting bioreactor at the desired sampling times. In some cases specified below, the results from two and three experimental runs were used to determine weight loss values and the standard deviation. All bioreactors were maintained at 53°C, 60% moisture content, and aerated at 100 mL/min. Abiotic reactors were used to assess whether chemical degradation of the polymeric materials occurs under the test conditions. A detailed description of the bioreactor configuration and the operational procedure have been published elsewhere [9, 10]. During the degradation testing, residual polymer films were recovered from the composting and abiotic test vessels at desired sampling time intervals. The recovered materials were cleaned by immersing in deionized water to remove the coarse particles on the film surface followed by gentle blotting with 70% isopropyl alcohol moistened Kimwipes EX-L (Kimberly-Clark, Fisher Scientific, Pennsylvania). The weight of residual film material was measured after the films were dried in a vacuum oven at 30°C (10 mmHg) to constant weight. Results of polymer film weight loss at a sampling time were expressed as the difference in weight between initial and sampling, normalized to the total surface area (two exposure sides for each film, units of $\mu g \cdot mm^{-2}$).

Synthetic Compost Mixture Formulation

Synthetic MSW (municipal solid waste) composting mixtures were made from various materials available naturally or commercially, and all materials were used without any treatment unless otherwise specified [12a]. All compost formulations used in this work are presented in Table 1. Tree leaves (approximately 1:1 ratio of oak and maple) collected in the fall of 1992 were air-dried and then stored at 4°C for later use. Shredded paper was a mix of computer paper with newspaper in a ratio of 1:1 (dimensions 2 mm × 25 mm). The food waste used, which contained 3.2% protein and 13.8% carbohydrates, was a commercially available mixture of frozen vegetables including carrots, green beans, green peas, corn, and lima beans. This vegetable mixture was kept frozen and thawed prior to use. The meat waste was simulated by mixing dog and cat food in a 1:1 ratio (w/w). The major source of sawdust was from maple wood. Steam-exploded wood was generated at log $R_0 = 3.7$, 202°C for 5.0 minutes on a mixture of maple and oak wood chips (Biobased Materials Research Center, Virginia Tech.). Dehydrated cow manure which contained a total nitrogen content of 2.0% (w/w) was from Earthgro, Inc. (Lebanon, Connecticut). Other materials used in this study included glass beads (2.0 mm in diameter, Fisher Scientific), urea (laboratory grade, Fisher Scientific), corn starch (Sigma), and Recycle compost inoculum (Ringer Compost Seeds, Minneapolis, Minnesota). The sources and nutritional content (as specified by the manufacturer) for Rabbit Choice, timothy and alfalfa, cat food, and dog food were obtained from BigRed (Syracuse, New York), Hay-Kob (L/M Animal Farms, Pleasant Plain, Ohio), Alpo (Lehigh Valley, Pennsylvania), and Cycle (The Quaker Oats Co., Chicago, Illinois), respectively.

Studies on the Percent Moisture in Composting Bioreactors [12a]

During simulation of different moisture conditions, the initial percent water content was obtained by first determining the percent water in the solid waste mixture by oven drying a sample at 105°C to constant weight (normally approximately 16 hours). Once the percent water in the solid waste mixtures were known, the mixtures were then amended with the required additional water so that 40, 50, and 60% (w/w) moisture values were obtained. Subsequently, moisture in the compost bioreactors was monitored by sampling the test solid waste mixtures at regular time intervals. Observed moisture losses were corrected by simply adding the required quantity of deionized water. In no case was it found that the percent moisture in the compost reactors increased during the composting test.

Elemental Analyses. Carbon and nitrogen elemental analyses on compost formulations were carried out by Atlantic Microlab, Inc. (Norcross, Georgia).

Respirometric Method to Measure Mineralization of Polymeric Materials in a Compost Environment [12b]

Respirometry Setup

The respirometric system used in assessing the mineralization of polymer carbon to CO_2 consisted of an air pump which served as an air source; an air pretreatment system to sterilize, remove CO_2, and control the flow rate; test vessels which contained predigested compost; and posttest vessel trapping of the CO_2 produced. The predigested compost was obtained by a 42-day maturation period of a defined synthetic waste mixture described in Ref. 12b. The waste mixture used was similar in composition to that of Formulation 2 given in Table 1. The matured or predigested compost supplied the indigenous microbial population and also served as a physiological and nutritional support. For each polymer sample analyzed, a total of six vessels were set up where three vessels were amended with 1.0 g of the polymer powder (particle size was between 300 and 400 μm using mesh screens) and the other three served as background controls to measure the basal CO_2 production from the indigenous microbial population. The test vessels were maintained at 53°C and 60% moisture throughout the analysis. The difference in CO_2 production between the test vessels with and without the polymer sample was attributed to the degradation of the polymer sample. Produced CO_2 was trapped by reaction with a NaOH solution. The amount of NaOH left unreacted was quantitated by titration with HCl to pH 9.0 using a Mettler DL 12 (Mettler Instrument Co., Highstown, New Jersey). A more detailed description of the methodology has been published elsewhere [12b].

RESULTS AND DISCUSSION

Laboratory-Scale Aerobic Composting Using Fresh Waste Mixtures to Study Plastic Degradation

Normalization of Polymer Film Weight Using the Initial Film Surface Area

The weight loss values for the exposed CA and cellophane films are reported normalized to the exposed initial surface area (including both sides of a film). In other words, the film weight loss values were calculated by measuring the gravimetric weight loss and dividing this by the initial film surface area (units of μg/mm^2). For a sample density of 1 g/mL, a 12.5-μg/mm^2 normalized film weight loss value corresponds to a loss in film thickness of approximately 0.0254 mm (1 mil). In carrying out this analysis of the film weight loss, it is assumed that the degradation processes involved (such as enzyme-catalyzed chain cleavage and subsequent dissolution of degradation products) occurs exclusively at the surface of the films and that the normalized weight loss value is independent of film thickness. This method was chosen in place of percent weight loss of plastic specimens since the latter measure does not take into account the importance of available surface area of exposed samples. However, it must be noted that the assumption that the normalized film weight loss values will be independent of film thickness is an idealized model. Problems such as changes in the film surface area due to film pitting and erosion as a function of time as well as the study of plastics that contain water-soluble components that freely diffuse from the specimen into the compost are expected to result in serious deviations from the model presented above.

Cellulose Acetate (CA) Normalized Film Weight Loss Measurements

The normalized weight loss values for CA DS-1.7 and cellophane films (film thickness values of ~0.025 and ~0.041 mm, respectively) are shown in Fig. 1. Compost Formulation 2 was used for this study. About 35% of the film weight loss

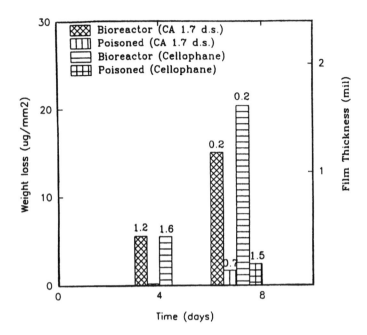

FIG. 1.   Normalized weight loss of CA DS-1.7 and cellophane films in Formulation 2. The standard deviation values are given above the bars.

for the two sample types was observed on Day 4 in the biologically active bioreactors, while there was almost no change for the films in the poisoned reactor. Both samples had completely disintegrated at the end of a 7-day exposure period and were no longer recoverable. Similar film weight loss values for 7-day exposure periods were observed for biologically-active test vessels which contained Formulation A (see below and Table 1). In contrast, the samples exposed in the poisoned control vessels (see Materials and Methods Section) showed comparatively small weight loss values (see Fig. 1). Therefore, the disappearance of these materials over the 7-day exposure period was due in large part to biologically-mediated processes.

CA of relatively higher DS was exposed to the compost bioreactors containing Formulation 1 (see Table 1). The weight loss of CA DS-2.5 films (film thickness, ~0.05 mm) was observed periodically over an 18-day exposure period (see Fig. 2). By the end of the 18-day exposure period, the weight loss was 23.5 ± 2 $\mu g/mm^2$, which corresponded to essentially complete disappearance of the polymer films. In the poisoned reactors, CA DS-2.5 films showed negligible weight loss over an 18-day exposure period. Once again, by comparison of the biologically-active and poisoned reactor results, it is concluded that the disappearance of CA DS-2.5 is due in large part to biologically-mediated processes. In addition, as would be expected, the CA DS-1.7 films showed much more rapid degradation kinetics relative to the 2.5-DS film samples.

Changes in the dry weight and pH of the compost as a function of time have been reported by us elsewhere [9]. Furthermore, residual CA polymers after various composting exposure periods have been characterized for changes in molecular

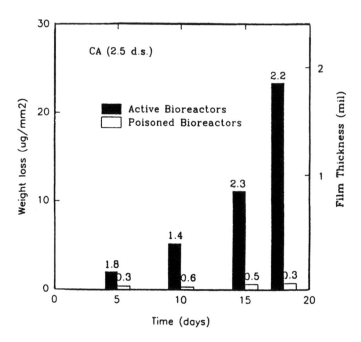

FIG. 2.   Normalized weight loss of CA DS-2.5 films in Formulation 1. The standard deviation values from one experiment (triplicate in bioreactors) are given above the bars.

weight and DS [11]. Colonization and degradation at the surface of exposed CA Samples was studied by SEM [11].

Effect of Compost Mixture Composition on Polymer Film Weight Loss.   Synthetic MSW mixtures were prepared using different component materials (see Table 1). Primarily CA DS-1.7 films and, in some cases, cellophane films were chosen as the polymer candidates for testing. The only difference between Formulations 1 and 2 is that the latter does not include glass beads, aluminum, and steel shavings (Table 1). Both mixes have almost identical C/N ratios (14.0 and 13.9, respectively). The extent of film weight loss as a function of the exposure time for CA DS-1.7 and cellophane films were compared for exposures to laboratory-scale reactors containing Formulations 1 and 2. Figure 3 shows that the weight loss of both film types exposed to composting reactors containing either Mixture 1 or 2 were virtually identical. Also, the time for complete film disappearance in these mixes was ~7 days for both film types.

Figure 4 shows the weight loss as a function of exposure time for CA DS-1.7 film exposed to other compost formulations. Mixture A is identical to Mixture 2 with the exception that steam-exploded wood was substituted for tree leaves in the formulation. The substitution of steam-exploded wood had no apparent effect on the time required for CA DS-1.7 film weight loss and subsequent disappearance (~7 days) although the C/N was 21.2, which is higher than Mix 2. This general

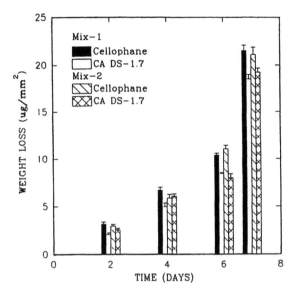

FIG. 3. Normalized weight loss of CA DS-1.7 and cellophane films in Compost Mixes 1 and 2. Vertical lines above the bars indicate the standard deviation values of three experiments.

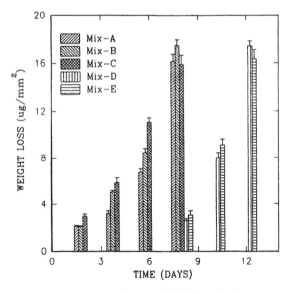

FIG. 4. Normalized weight loss of CA DS-1.7 films in bioreactors containing mixes A, B, C, D, and E. The vertical lines above the bars indicate the standard deviation values of two experiments.

lack of dependence of the mixture composition on the time required for CA DS-1.7 film weight loss and subsequent disappearance was also observed for Mixtures B and C relative to Mixtures 1, 2, and A (see Figs. 3 and 4 and Table 1). Mixtures B and C replace food and meat wastes, sawdust, and urea with the materials dried timothy and alfalfa, respectively. The nutritional contents for Mixtures B and C components timothy and alfalfa are shown elsewhere [12a]. The C/N ratio was 61.4 and 31.2 for Mixtures B and C, respectively. The artificial waste Mixtures B and C are exciting alternatives since they contain component materials that are readily available from commercial sources having consistent nutritional contents. These formulations offer the opportunity to carry out laboratory-scale compost testing with excellent run-to-run and laboratory-to-laboratory reproducibility.

Further study of the effects of mix composition on CA DS-1.7 film weight loss was carried out by investigation of Formulations D and E as synthetic waste recipes for laboratory-scale composting. Formulation D was obtained from C by the omission of cow manure and dried alfalfa and addition of rabbit choice. Formulation of mixture E was obtained from C by the omission of dried alfalfa and the addition of starch and urea (see Table 1). The C/N ratio of Formulations D and E are 49.9 and 24.7, respectively. Interestingly, the exposure time required for CA DS-1.7 film weight loss and subsequent film disappearance in the biological active reactors containing Formulations D and E was extended to 12 days relative to 7 days as was observed for Mixtures 1, 2, A, B, and C (see Figs. 3 and 4). This variation of time required for CA DS-1.7 film weight loss may be due to the differences in physical form and nutritional contents of Formulations D and E [14]. The relatively slower film weight loss in Formulations D and E cannot simply be attributed to the mixture C/N values since the values for C/N for these mixtures were within the upper (61.4, Mix B) and lower (13.9, Mix 2) C/N values of the other formulations studied. It should be pointed out that in control poisoned reactors (see Materials and Methods Section) constructed using all of the formulations described in Table 1, no significant weight loss of either CA DS-1.7 or cellophane polymer films was observed for up to 12-days exposures. Therefore, the observed film weight loss must at least in part be attributed to biomediated processes.

As expected, the results above provide evidence that the rate of film weight loss is dependent on the compost formulation. However, the similarity of film weight loss rates between Mixes 1, 2, A, B, and C was surprising. It should be noted that a biodegradable polymer film placed in a compost reactor is simply an alternative carbon source for microorganisms which are present in the composting reactor. The rate of biodegradation of the exposed polymer relative to the utilization of other nutrients in the mixture will, of course, be dependent on whether the polymer serves as a preferred carbon source. It may be speculated that exposure of another polymer type which has a significantly different nutrient value to compost microorganisms than CA DS-1.7 might show a much larger variance in film weight loss rates as a function of the compost formulation used. The fact that CA DS-1.7 showed rapid film weight loss rates in all of the mixtures studied serves as important evidence that CA DS-1.7 is in fact a readily biodegradable polymer that is utilized competitively with other compost nutrients. This is further supported by the similarity of the time dependence of film weight loss for CA DS-1.7 and cellophane films exposed to Formulations 1 and 2.

Effect of Compost Moisture Content on Polymer Film Degradation

Compost Mixture 2 (see Table 1) was chosen for analysis of the effect of compost moisture content on the rate of polymer film degradation. In this study, all other environmental parameters in the laboratory-scale composting experiment, such as the rate of aeration, C/N ratio, and the incubation temperature (see Materials and Methods Section and Refs. 9–11), were kept constant. It was determined that for moisture contents of 60, 50, and 40% the time for complete CA DS-1.7 and cellophane (data not shown) film disappearance was 6, 16, and 30 days, respectively (Fig. 5). During the same period of time, films in the control reactors showed little to no weight loss (<5%) due to chemical or physical erosion. Therefore, the rate of film weight loss is extremely sensitive to the compost percent moisture. This is consistent with the fact that the availability of water is a critical factor dictating microbial viability and metabolic activity [13].

Mineralization of Cellulose Acetates

Mineralization of a polymeric material by microorganisms under aerobic conditions produces CO_2, H_2O, microbial biomass, and humus. Therefore, a respirometric technique was used to quantitate the evolved CO_2 due to polymer biodegradation. The method was summarized above (see Respirometry Setup Section) and was described in detail elsewhere [12b]. The low basal level of CO_2 production in test vessels which did not contain the polymer substrate (approximately 12–15 mg CO_2-C per day) was found to be linear as a function of time [11, 12b] and was

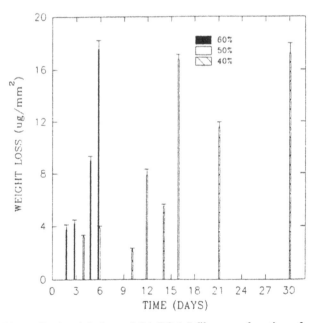

FIG. 5.   Normalized weight loss of CA DS-1.7 films as a function of exposure time in laboratory-scale compost reactors at moisture contents of 60, 50, and 40%. The vertical lines above the bars indicate the standard deviation values of three experiments.

subtracted from that produced in polymer-amended vessels to obtain the total CO_2 and the corresponding percentage of the theoretical CO_2 formed as a results of substrate mineralization (see Figs. 6 and 7). Observation of Fig. 6 shows that exposure of CA DS-1.7 to the predigested compost matrix produced 72.4% of its theoretical carbon as CO_2 after a total incubation period of 24 days. In comparison, the total time required for CA DS-2.5 to reach the plateau value of 77.6% substrate carbon conversion to CO_2 was 60 days (see Fig. 7). Of additional interest, the rate of CA DS-1.7 mineralization increased rapidly after only a 10-day lag period whereas the lag period prior to rapid CA DS-2.5 mineralization was considerably longer (approximately 25 days). In agreement with these results, both the lag period prior to film weight loss and the total time for film disappearance were notably longer for CA DS-2.5 relatively to CA DS-1.7 (see Ref. 9 and above). Indeed, when one considers probable mechanisms for CA biodegradability (see below), it is not at all surprising that CA of relatively higher DS shows relatively slower rates of biodegradation. However, most importantly, the results shown in Fig. 7 suggest that CA with a degree of acetylation up to 2.5 is mineralized to a high degree under suitable environmental exposure conditions.

CA Biodegradation Mechanistic Considerations

A hypothesis has previously been presented by us [11] that deacetylase enzymes are required to provide chains or chain segments with DS values of less than approximately 0.7. Once this critical degree of deacetylation is reached, cellulase

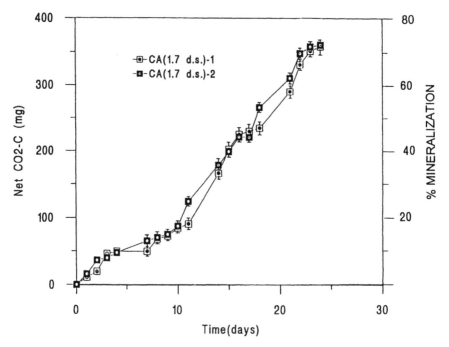

FIG. 6.   Respirometric mineralization of powdered cellulose acetate (DS-1.7) in a matured synthetic MSW compost matrix under aerobic thermophilic conditions.

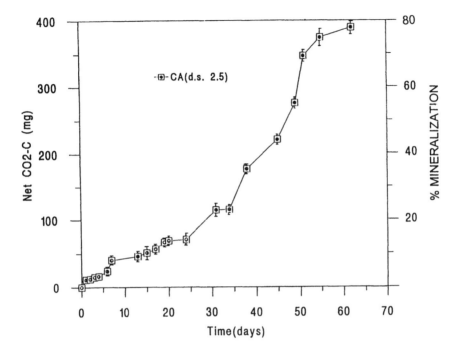

FIG. 7. Respirometric mineralization of powdered cellulose acetate (DS-2.5) in a matured synthetic MSW compost matrix under aerobic thermophilic conditions.

enzymes cause events of chain cleavage which are accelerated by further chain deacetylation. It should be noted that the critical value of deacetylation required for glycosidic bond cleavage by cellulase enzymes has not as yet been firmly established for a range of cellulase enzyme systems.

SUMMARY OF RESULTS

In-laboratory simulation of a synthetic MSW aerobic compost process using fresh synthetic waste formulations was developed and used to study the biodegradability of CA films. The exposure of CA (DS-1.7 and -2.5) and cellophane films to biologically-active and poisoned in-laboratory composting test vessels clearly demonstrated that these materials were biodegradable under controlled composting conditions. In other words, the film weight loss observed resulted, at least in part, from biologically-mediated processes.

The conversion of CA DS-1.7 and DS-2.5 to CO_2 was monitored by respirometry where powdered CA was placed in a predigested compost matrix. A lag phase of 10- and 25-day duration for CA DS-1.7 and -2.5, respectively, was observed after which the rate of degradation increased rapidly. Mineralization of exposed CA DS-1.7 and -2.5 powders reported as the percent theoretical CO_2 recovered reached 72.4 and 77.6% in 24 and 60 days, respectively. The results of this study demonstrated that microbial degradation of CA films exposed to aerobic thermophilic simulated composting conditions not only results in film weight loss but also conversions to greater than 70% of the theoretical recovered CO_2 for CA (1.7 and 2.5 DS) substrates. Thus, it appears that CA with DS $\leq$ 2.5 can undergo high extents of mineralization when exposed to the appropriate disposal conditions.

Alternative synthetic compost mixture formulations for laboratory-scale degradation testing were investigated. It was found that rapid CA DS-1.7 film weight loss and ultimate film disappearance was achieved with a variety of compost formulations. In two of the seven compost mixture formulations investigated, the extent of CA DS-1.7 film weight loss as a function of time was considerably slower. Thus, a dependence of film weight loss on the compost mixture composition was demonstrated. However, the fact that film weight loss occurred to a similar extent as a function of exposure time for five of the seven mixtures studied was surprising and indicates that CA DS-1.7 is a readily utilized carbon source relative to other available organic matter in the compost formulations studied. An important conclusion of this work is that laboratory-scale composting tests can be carried out using readily available commercial materials to construct synthetic MSW mixtures. It is expected that with the adoption of both standard synthetic waste mixtures as well as testing protocols, reproducibility of test results will be extended beyond run-to-run so that reproducibility between laboratories will ultimately be achieved.

Moisture conditions in composting runs was found to have a profound affect on the extent of CA DS-1.7 film weight loss as a function of the exposure time. Thus, the percent moisture is an extremely sensitive compost test parameter that must be carefully controlled to achieve rapid degradation of organic matter as well as reproducible polymer degradation testing results.

ACKNOWLEDGMENTS

We are grateful for the financial support of this research by the NSF Biodegradable Polymer Research Center (BPRC) at the University of Massachusetts Lowell and the Division of Environmental Engineering and Pollution Control, 3M Company.

REFERENCES

[1] H. Brandl, R. A. Gross, R. W. Lenz, and R. C. Fuller, *Adv. Biochem. Eng., 41*, 77–93 (1990).
[2] D. L. Kaplan, J. M. Mayer, D. Ball, J. McCassie, A. L. Allen, and P. Stenhouse, in *Biodegradable Polymers and Packaging Materials* (C. Ching, D. Kaplan, and E. Thomas, Eds.), Technomic Co., Lancaster, Pennsylvania, 1993, pp. 1–42.
[3] A. Gibbons, *Technol. Rev., 92*, 69–73 (1989).
[4] J. Glenn and N. Goldstein, *BioCycle, 33*, 48–52 (1992).
[5] N. Goldstein and R. Stevteville, *Ibid., 33*, 44–47 (1992).
[6] M. S. Finstein, in *Environmental Microbiology* (R. Mitchell, Ed.), Wiley-Liss, New York, 1992, pp. 355–374.

[7] R. L. Chaney, in *The BioCycle Guide to the Art and Science of Composting*, JG Press, Emmaus, Pennsylvania, 1991, pp. 240–253.

[8] J. A. Hogan, F. C. Miller, and M. S. Finstein, *Appl. Environ. Microbiol.*, 55, 1082–1092 (1989).

[9] J.-D. Gu, D. Eberiel, S. P. McCarthy, and R. A. Gross, *J. Environ. Polym. Degrad.*, 1, 143–153 (1993).

[10] R. A. Gross, J.-D. Gu, D. T. Eberiel, M. Nelson, and S. P. McCarthy, in *Biodegradable Polymers and Packaging Materials* (C. Ching, D. Kaplan, and E. Thomas, Eds.), Technomic Co., Lancaster, Pennsylvania, 1993, pp. 257–279.

[11] J.-D. Gu, D. Eberiel, S. P. McCarthy, and R. A. Gross, *J. Environ. Polym. Degrad.*, 1(4), 281–291 (1993).

[12] (a) J.-D. Gu, S. Yang, R. Welton, D. Eberiel, S. P. McCarthy, and R. A. Gross, *Ibid.*, 2(2), 129–135 (1994). (b) J.-D. Gu, S. Coulter, D. Eberiel, S. P. McCarthy, and R. A. Gross, "A Respirometric Method to Measure Mineralization of Polymeric Materials in a Matured Compost Environment," *Ibid.*, In Press.

[13] F. C. Miller, *Microbiol. Ecol.*, 18, 59–71 (1989).

[14] K. Nakasaki, M. Sasaki, M. Shoda, and H. Kubota, *Appl. Environ. Microbiol.*, 49, 724–726 (1985).

[15] K. Nakasaki, M. Shoda, and H. Kubota, *J. Ferment. Technol.*, 63, 537–543 (1985).

[16] K. Nakasaki, M. Shoda, and H. Kubota, *Ibid.*, 64, 539–544 (1986).

[17] P. D. Bach, K. Nakasaki, M. Shoda, and H. Kubota, *Ibid.*, 65, 199–209 (1987).

[18] F. C. Miller and M. S. Finstein, *J. Water Pollut. Control Fed.*, 57, 122–127 (1985).

[19] M. S. Finstein, F. C. Miller, and P. F. Strom, *Ibid.*, 58, 272–278 (1986).

[20] F. C. Miller, S. T. MacGregor, K. M. Psarianos, J. Cirello, and M. S. Finstein, *Ibid.*, 54, 111–113 (1982).

[21] S. T. MacGregor, F. C. Miller, K. M. Psarianos, and M. S. Finstein, *Appl. Environ. Microbiol.*, 41, 1321–1330 (1981).

[22] P. M. Strom, *Ibid.*, 50, 899–905 (1985).

[23] D. F. Gilmore, S. Antoun, R. W. Lenz, S. Goodwin, R. Austin, and R. C. Fuller, *J. Ind. Microbiol.*, 10, 199–206 (1992).

[24] A. M. Fogarty and O. H. Tuovinen, *Microbiol. Rev.*, 55, 225–233 (1991).

[25] D. R. Reinhart and F. G. Pohland, *J. Ind. Microbiol.*, 8, 193–200 (1991).

[26] US Food and Drug Administration, *Environmental Assessment Technical Assistance Handbook PB87-175345*, National Technical Information Service, Washington, D.C., 1987.

[27] J. M. Mayer, M. Greenberger, D. L. Kaplan, R. A. Gross, and S. P. McCarthy, *Polym. Mater. Sci. Eng.*, 63, 858–861 (1990).

[28] J. E. McCassie, J. M. Mayer, R. E. Stote, A. E. Shupe, P. J. Stenhouse, P. A. Dell, and D. L. Kaplan, *Ibid.*, 67, 353–354 (1992).

[29] J. M. Mayer and D. Kaplan, in *Biodegradable Polymers and Packaging Materials* (C. Ching, D. Kaplan, and E. Thomas, Eds.), Technomic Co., Lancaster, Pennsylvania, 1993, pp. 233–245.

[30] J. E. McCassie, J. M. Mayer, R. E. Stote, A. E. Shupe, P. J. Stenhouse, P. A. Dell, and D. L. Kaplan, *Ibid.*, pp. 247–256.
[31] R. Tillinger, B. De Wilde, and L. De Baere, *Polym. Mater. Sci. Eng., 67,* 359–360 (1992).
[32] L. De Baere, in *Biotechnology and Bioengineering Symposium N*, Wiley, New York, 1986, p. 321.

APPLICATIONS AND ENVIRONMENTAL ASPECTS OF CHITIN AND CHITOSAN

MARTIN G. PETER†

Institut für Organische Chemie und Biochemie der Universität
Gerhard-Domagk-Str. 1, D-53121 Bonn, Germany

ABSTRACT

A survey on the properties of the natural nitrogen-containing poly-saccharides chitin and chitosan is given. Outstanding features are the materials' mechanical and chemical properties which offer numerous, largely unexplored applications in technology, chemistry, medicine, and agriculture. Derivatives of chitin and chitosan are accessible by reactions of the hydroxy and amino groups with appropriate reagents. Various types of gels, membranes, and fibers, including polycationic and water-soluble materials, can be formed. Production of chitin and chitosan from waste crab shells involves environmentally safe processes. The polysaccharides are recycled in nature by enzymatic degradation and reuse of N-acetylglucosamine for biosynthesis and catabolism. Research deficits exist in economically competitive production technologies, construction of composite materials, and the ecological aspects of chemically modified chitosan.

†Present address: Institut für Organische Chemie und Strukturanalytik, Universität Potsdam, Am Neuen Palais 10, D-14469 Potsdam, Germany. Telephone: (0331) 977-1450/ 1444. FAX: (0331)977-1131. e-mail: peter@serv.chem.uni-potsdam.de.

37

INTRODUCTION

Chitin is, after cellulose, the second most abundant polysaccharide on earth. It consists of β-1,4-linked glucosamine (GlcN) with a high degree of N-acetylation. The M_R of natural chitin is estimated to be in the range of $1-2 \times 10^6$ daltons. When the polysaccharide contains lower proportions of N-acetyl groups (i.e., GlcNAc units), it is named chitosan. The criteria for distinguishing between chitin and chitosan are the solubilities of the polymers in dilute aqueous acid: chitin is insoluble while chitosan forms viscous solutions. As a rule of thumb, the degree of N-acetylation (DA) of chitosan is ≤ 40. Chemical derivatives of chitin or chitosan are named by using the kind of functional group introduced as a prefix (e.g., carboxymethyl chitosan) or as a suffix (e.g., chitosan sulfate). A comprehensive treatise of general "chitinology" is available in Muzzarelli's unsurpassed classical textbook [1], while a more recent book deals more specifically with chitin chemistry [2]. The growing interest in chitin and chitosan is reflected by a series of International Chitin Conferences and the published proceedings [3-5].

OCCURRENCE AND RESOURCES

Chitin occurs as a structural component of the exoskeleton of insects and crustacea as well as in the cell wall of yeasts and fungi where its relative amounts are in the range of 30 to 60%. According to earlier estimates, the resources of chitin in marine organisms are in the order of 10^6 to 10^7 tons. The annual production of chitin in zooplankton alone is estimated to be several billion tons. Actually, as Yu et al. have stated in a pictorial formulation, there is a constant "rain" of chitin on the ocean floor [6]. Chitosan occurs naturally in several fungi, especially *Mucor* species.

FUNCTIONS IN NATURE

Chitin serves as a fibrous strengthening element in biological composite materials. Thus, it is nearly always associated with proteins which function as the matrix for enforcing reactions such as phenolic tanning in the exoskeleton of insects [7] and/or mineralization in the shell of crustaceae, some flies, and nacre (mother of pearl) [8].

BIOSYNTHESIS

The biosynthesis of chitin takes place in the membrane-bound protein complex chitin synthase. Chain elongation occurs by sequential transfer of GlcNAc from UDP-GlcNAc to the nonreducing end of the growing polymer. In contrast to the situation in arthropods, a detailed picture of chitin biosynthesis has been established in the yeast *Saccharomyces cerevisiae* [9]. A detailed description of chitin biosynthesis is beyond the scope of this article, and the reader is referred to some recent reviews [9, 10].

BIODEGRADATION

The natural pathway of chitin metabolism includes enzyme-catalyzed hydrolysis by chitinases. There are various forms of these enzymes which are usually grouped into endo- and exo-enzymes (see Fig. 1). The chitinases (EC 3.2.1.14) cleave the polysaccharide and higher chitooligosaccharides in a random fashion to give N,N'-diacetylchitobiose and N,N',N''-triacetylchitotriose as the final products. There are also microbial chitinases that cut off chitobiose from the nonreducing end of oligosaccharides [11]. Further degradation of chitobiose takes place by N-acetylglucosaminidases (EC 3.2.1.30), formerly named chitobiases. The nomenclature of chitinases has been critically discussed by Muzzarelli [12]. Recently, a chitodextrinase was identified in the marine bacterium *Vibrio furnissii* by Bassler et al. [13]. This enzyme does not hydrolyze chitin, chitobiose, or chitotriose, but cleaves readily the tetra-, penta-, and hexamer, yielding exclusively the tri- and disaccharides. Most interestingly, the bacteria show chemotaxis toward chitooligosaccharides.

Another pathway of chitin turnover occurs in fungi which deacetylate chitin to chitosan by the action of chitin deacetylase (EC 3.5.1.41) [14, 15]. Undoubtedly, this reaction plays an important role in the biosynthesis of chitosan. Chitosan itself is degraded by chitosanes (EC 3.2.1.99) that occur in microorganisms and, after elicitation by chitooligosaccharides, as phytoalexins in plants.

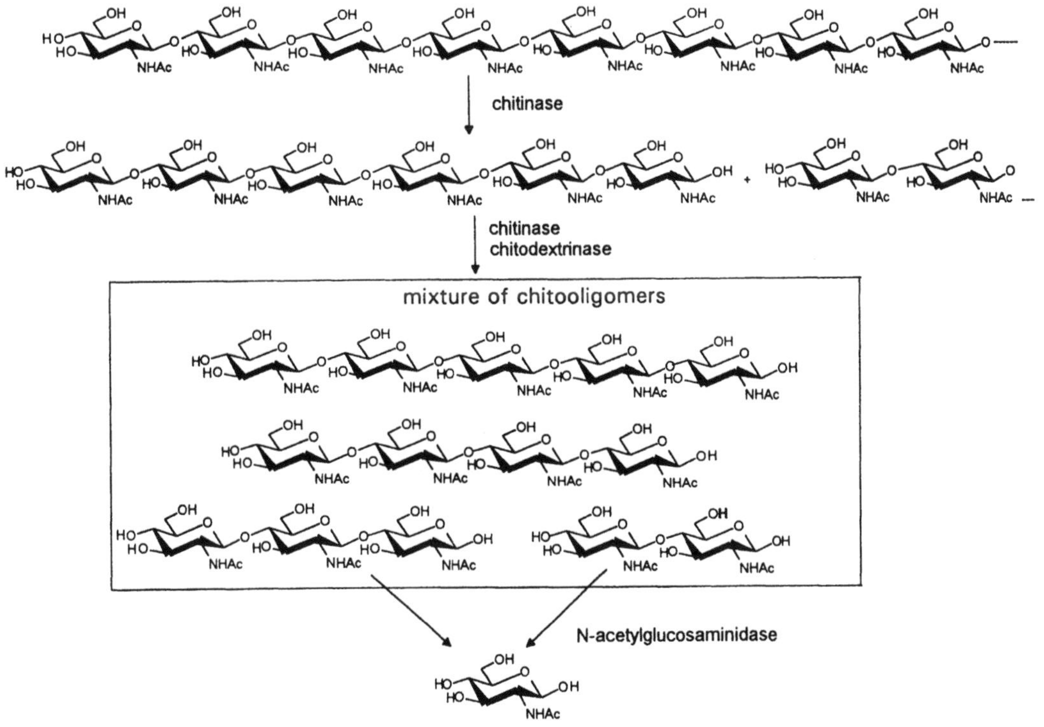

FIG. 1.   Enzymatic degradation of chitin.

Chitin and chitosan are also degraded by various lysozymes (EC 3.2.1.17) which are widely distributed in plants and animals. A residual number of acetyl groups ($DA \geq 16$) and the free hydroxy group at C-3 of the sugar units are required in order to observe reasonable rates of hydrolysis [16].

PRODUCTION OF CHITIN AND CHITOSAN

Production of chitin by chemical synthesis is technically not possible though small quantities of defined chitooligomers are principally accessible by appropriate procedures. Also, biotechnological production of chitin is presently not economically attractive. The primary sources of chitin are therefore the tremendous amounts of waste crab and krill shells from the fishing industry, while chitosan is usually produced by chemical deacetylation of chitin, though fermentation processes are also possible. In order to obtain rather crude chitin, crab shells are decalcified at ambient temperature by means of dilute aqueous hydrochloric acid, followed by extensive washing with water and deproteination with dilute sodium hydroxide. Pigments, such as carotenoids, may be extracted with appropriate organic solvents. Further purification is achieved by adding an ice-cold solution of chitin in 12 N hydrochloric acid slowly to a vigorously stirred large volume of water. This procedure may be repeated several times. For chitosan production, crude chitin is deacetylated with 40–50% sodium hydroxide at 110–115°C. All procedures involving acid or heat lead to a more or less extensive depolymerization which must be controlled by temperature, reaction times, and proper reagent compositions. Commercial preparations have M_r values between 10^4 and 10^5 daltons, though higher molecular weight materials are available also. A comprehensive and critical evaluation of chitin and chitosan processing was given by Muzzarelli [1].

The procedures employed in chitin/chitosan production require relatively harmless chemicals. The acetyl groups are recovered in the form of sodium acetate. Calcium carbonate, which is another major component of crab shells, is converted to calcium oxide and sodium carbonate. In addition, pigments such as astaxanthin may be recovered as high value side products. Thus, from the ecological view, chitin and chitosan production bears fewer problems than the production of cellulose. The market value of chitin is in the order of DM 10 to DM 300 per kg, depending on the product specification. High purity, pyrogen-free chitosan for medical applications may cost as much as DM 70,000 per kg.

MATERIALS PROPERTIES

The quaternary structure of chitin is similar to the structure of cellulose. The chitin of insect and crustacean cuticle occurs in the form of microfibrils typically 10–25 nm in diameter and 2–3 μm in length. Three natural modifications are known which differ in the orientation of the polysaccharide chains within the microfibrillae, namely, α- (antiparallel), β- (two parallel, one antiparallel), and γ- (random orientation) chitin. The most abundant form is α-chitin. Young's modulus of elasticity of chitin fibrils of locust tendon ($E = 70$–90 GPa) is comparable with that of gold.

CHEMICAL PROPERTIES OF CHITIN AND CHITOSAN

Chitin may be converted by use of proper reagents into a number of O-alkyl and O-acyl derivatives, as has been summarized in a recent review by Hirano [17] (see Fig. 2). Chitosan behaves as a moderately basic ($pK_a = 6.3$; cf. GlcN: $pK_a = 7.47$) cationic polyelectrolyte forming salts with acids. This is definitely an advantage in comparison with cellulose which, in order to achieve ion-exchange properties, has to be chemically converted into the appropriate amino group-containing derivatives. In addition, the presence of the primary amino group in chitosan offers further possibilities for modifications such as N-acylation, N-alkylation, and N-alkylidenation (see Fig. 2). Controlled degradation of chitin either by enzymatic or by acid-catalyzed hydrolysis gives chito-oligosaccharides which, besides possessing interesting biological activities, are valuable starting materials for the synthesis of oligosaccharides and other derivatives [18].

USES OF CHITIN AND CHITOSAN

The worldwide research activities going on in the chitin field are reflected by the number of patents, including patent applications, and other publications. Table 1 gives a survey of the results of a search in *Chemical Abstracts* on chitin-related sources which were recorded in the year 1993. Though this gives only a very rough picture which does not tell anything about the subjects of research, it is immediately clear that most work on chitin and chitosan is performed in Japan. Actually, this correlates with the annual production rate of presently ca. 3000 tons of chitin in Japan. The major uses of chitin and chitosan are in the food and cosmetics industries (S. Hirano, personal communication, May 1993). The rationale for applications in those areas are the formidable gel-forming properties as well as the moistening and stiffening effect of chitosan and certain derivatives, such as hydroxyalkylchitosans which are essentially nontoxic. Other applications which are based on the binding and flocculation of proteins and on the chelation of various heavy metal ions (see Table 2) are realized in the treatment of wastewaters. A summary of chitin and chitosan uses in Japan is given in Table 3 [19].

POTENTIAL FOR FURTHER USES

The fascinating properties of chitin, chitosan, and derivatives have initiated numerous research activities on the industrial as well as on the academic level, which result in an ever-increasing amount of new potential applications. It is impossible here to give a comprehensive treatment of the subject, and reference is given to the monographs [1, 2] and conference proceedings [3–5], as well as compilation of data published recently [20]. Just a few highlights will be mentioned.

Membranes formed from chitosan are less permeable for oxygen, nitrogen, and carbon dioxide than cellulose acetate membranes. This, in combination with mechanical strength, offers superior wrapping and packing materials in the food industry. Chitosan can be used as a film-forming product for fruit and seed conservation. The procedure is rather simple: the seeds or fruits are dipped into a 2–3%

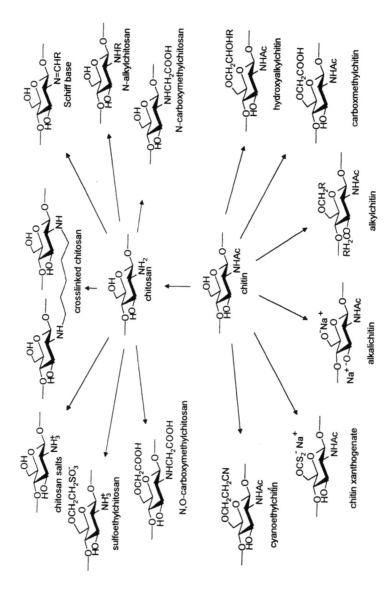

FIG. 2.   Chemical derivatization of chitin and chitosan.

TABLE 1.   Worldwide Research Activities 1993 in the Fields of Chitin and Chitosan as Reflected by the Number of Publications According to *Chemical Abstracts* Citations

Country	Patents	Other publications	Total
Japan	126	88	214
USA	25	59	84
UK	4	19	23
Germany	7	15	22
France	9	11	20
Canada	2	13	15
People's Republic of China	2	13	15
Russia	2	12	14
Italy	5	8	13
India		9	9
South Korea		7	7
Israel	2	4	6
Belgium		5	5
Netherlands		5	5
Australia	1	3	4
Cuba		4	4
Poland		4	4
Spain		4	4
Switzerland	2	2	4
Denmark		3	3
Greece		3	3
Sweden	1	2	3
Turkey		3	3
Brazil		2	2
Singapore		2	2
Slovenia		2	2
Argentina		1	1
Chile		1	1
Czech Republic		1	1
Egypt		1	1
Indonesia		1	1
New Zealand		1	1
Norway		1	1
Romania		1	1
Slovakia		1	1
Total	190	314	501

TABLE 2. Compilation of Data from Various References on the Binding Capacities of Polymers for Metal Ions (meq·g⁻¹)[a]

Polymer	Method[b]	Mg^{2+}	Ca^{2+}	Sr^{2+}	Ba^{2+}	Mn^{2+}	Ni^{2+}	Cu^{2+}	Cd^{2+}	Pb^{2+}	Zn^{2+}	Hg^{2+}	Co^{2+}	Cr^{2+}	Fe^{2+}	Ag^+	Au^{3+}	Pt^{4+}	Pd^{2+}
Chitosan 45%DA	A	0.3	0.8	1.5	1.1	1.1	3.5	5.3	6.5		5.5								
Chitosan 97%DA	A	0.5	0.4	0.6	0.8	0.5	2.3	4.8	4.9		3.2								
Chitosan	B					1.44	3.15	3.12	2.78	3.97	3.70	5.60	2.47	0.46	1.18	3.26	5.84	4.52	6.28
p-Aminostyrene	B						0.02	1.31	0.10	0.17	0.52	5.70	0.07	0.03	0.05	1.98	5.75	3.82	0.87
6-O-CM-chitin[c]	C	2.08	2.50	2.12	2.15	2.22	2.22		2.42	2.13									
3,6-Di-O-CM-chitin[d]	C	4.0	3.3[a]	2.8	10.1↓	.34	4.1↓	3.1↓	9.2↓	8.5↓									

[a]↓: Gel formation or precipitation.
[b]Method: A: 0.04 M metal ion, pH 7.4, 20 h, 30°C; B: 1 g polymer, 0.02–0.4 mM metal chloride or nitrate, 24 h; C: not specified.
[c]6-O-CM-chitin: 6-O-carboxymethylchitin.
[d]3,6-Di-O-CM-chitin: 3,6-di-O-carboxymethylchitin.
[e]Ca^{2+} can be desorbed by means of EDTA.

TABLE 3. Practical Applications of Chitin-Derived Products in Japan (1988) [19]

Area	Chitin-based product	Application
Medicine	Chitin Chitin fiber Chitosan-collagen composite	Wound dressings Wound sutures Artificial skin
Cosmetics	Chitosan Hydroxyalkyl chitosan Liquid Chitin	Creams and other skin-care products Hair stiffeners
Technical applications	Chitin membranes Depolymerized chitosan Schiff bases	Loudspeakers Adsorption of endotoxins and nucleic acids Matrix for electron microscopy
Environmental	Chitosan and chitosan salts	Flocculant for purification of protein-containing wastewaters
Agriculture	Chitosan	Compostation accelerator
Food technology	Chitosan	Food additives
Biotechnology	Chitosan	Porous particles for bioreactors Immobilization of cells and enzymes
Chemistry	Chitosan Alkali chitin	Carrier for catalysts Intermediate for the synthesis of chitin derivatives

solution of chitosan lactate. The conserving effect is due in part to the antimicrobial properties of the polysaccharide.

The presence of hydroxy and amino groups of chitosan allows for a wide variety of chemical modifications including reactions with crosslinking agents. Thus, porous beads are made from crosslinked chitosan that are excellent carrier supports for the immobilization of enzymes and cells in biotechnology. Porous beads are also required for chromatographic stationary phases. A product, Chito-pearl, is commercially available in Japan. It has been used for the separation of proteins, peptides, and oligosaccharides.

Numerous medical applications have been suggested. Since chitin and chitosan can be degraded by lysozyme, they are useful as drug carriers and slow release formulations. Chitooligosaccharides exert antimicrobial and macrophage activating properties which are manifested in remarkable wound-healing acceleration by chitosan-based wound dressings. In fact, a chitosan-based product is registered and produced in Poland for applications in veterinary medicine.

Glucosamine, as the most abundant and cheapest amino sugar, is an excellent source from the chiral pool that can easily be used as starting material for the preparation of high value fine chemicals and enantiopure building blocks, such as amino alcohols, aldehydes, and acids, which thus are ultimatively derived from chitin [21]. The application of chitin as a source for the production of precious natural products has not yet been fully explored.

The design of chitin-based composite materials is still in its infancy. Most of the work done so far has focused on chitin/chitosan composites with cellulose. Chitosan membranes can be subjected to cold stretching, and fibers may be spun into a coagulation bath. Reacetylation yields chitin membranes and fibers. Numerous composite materials could principally be constructed by imitation of various natural biomaterials. This will certainly be one of many highly attractive research areas in the future.

RECYCLING

Any process of recycling should avoid consuming thermal energy from fossil resources for the degradation or transformation of materials, in particular when they have been generated previously at the cost of such resources. Recycling in nature always occurs in metabolic processes at moderate temperatures, converting one form of chemical energy into another, maintaining a strictly regulated balance between heat production in catabolic processes and high energy intermediates for biosynthetic requirements in anabolism. In the life cycle of arthropods, most of the huge amounts of chitin are recycled by enzymatic digestion during the molting process while the remainders are utilized after individual death by microorganisms for catabolism to GlcNAc-6-phosphate, fructose-6-phosphate, acetate, and ammonia and thus, ultimately, combustion to carbon dioxide and water.

Procedures for the recycling of chitosan membranes, wrappings, etc. have not yet been investigated. Since, in general, it is probably neither of economical nor of ecological advantage to recover intact chitin from composite materials, recycling of chitin or chitosan itself in an approach similar to that pursued with paper or with conventional plastics, i.e., shredding and remolding, appears impractical. Thus, the production of chitin will continue to rely on waste from seafood processing while the disposal of technical waste products can reasonably be envisaged to follow the way of microbial degradation, i.e., composting.

Rather little is known about the biodegradation of composite materials made from chitin and chemically modified chitosan. A recent study allows the conclusion that partially N-acylated derivatives of chitosan, O-carboxymethyl, and O-glycolylchitosans are degraded by the chitosanase of *Bacillus pumilus* [22]. Furthermore, there are a variety of fungi capable of using polyphenolic compounds as a source of carbon in peroxidative pathways. Thus, the chances for environmentally safe disposal of chitinous waste materials are excellent. However, this topic clearly needs further detailed research.

CONCLUSIONS AND FUTURE PROSPECTS

The majority of the applications are still in the developing stage. Major obstacles for large-scale technical applications of high quality chitin/chitosan are the high prices of the polymers and the difficulties in preparing uniformly reproducible charges in bulk quantities from various marine organisms. Issues for future research are:

- Development of economically competitive production processes
- Development of technical processes for the preparation of chitin derivatives and composite materials
- Investigations of the biodegradation of chemically modified chitosan

The situation of a rapidly rising demand for chitin and chitin based products may arise in the foreseeable future. Care must be taken in the exploration of natural resources, and concern about ecological questions could arise once chitin is produced from crabs and other marine organisms that are caught for the purpose of chitin production only. Therefore, the planning of an eventually large-scale utilization of chitin must also include thoughts about a possible decrease of biodiversity and shifts in the ecological balance in the marine environment.

REFERENCES

[1]   R. A. A. Muzzarelli, *Chitin*, Pergamon Press, Oxford, 1977.
[2]   G. A. F. Roberts, *Chitin Chemistry*, Mcamillan, Houndmills, 1992.
[3]   R. Muzzarelli, C. Jeuniaux, and G. W. Gooday (Eds.), *Chitin in Nature and Technology*, Plenum Press, New York, 1986.
[4]   G. Skjåk-Bræk, T. Anthonsen, and P. Sandford (Eds.), *Chitin and Chitosan*, Elsevier Applied Science, London, 1989.
[5]   C. J. Brine, P. A. Sandford, and J. P. Zikakis (Eds.), *Advances in Chitin and Chitosan*, Elsevier Applied Science, London, 1992.
[6]   C. Yu, A. M. Lee, B. L. Bassler, and S. Roseman, *J. Biol. Chem., 266*, 24260 (1991).
[7]   M. G. Peter, *Chem. Unserer Zeit, 27*, 189 (1993).
[8]   J. F. V. Vincent, *Structural Biomaterials*, Princeton University Press, Princeton, New Jersey, 1990, pp. 140, 169.
[9]   E. Cabib, *Adv. Enzymol., 59*, 59 (1987).
[10] E. P. Marks and G. B. Ward, in *Chitin and Benzoylphenylureas* (J. E. Wright and A. Retnakaran, Eds.), Dr. W. Junk Publishers, Dordrecht, 1987, p. 33.
[11] H. Diekman, A. Tschech, and H. Plattner, in Ref. 4, p. 207.
[12] R. A. A. Muzzarelli, *Chitin Enzymology*, European Chitin Society, Lyon and Ancona, 1993, p. VII.

[13] B. L. Bassler, C. Yu, Y. C. Lee, and S. Roseman, *J. Biol. Chem., 266*, 24276 (1991).

[14] L. L. Davis and S. Bartnicki-Garcia, *Biochemistry, 23*, 1065 (1984).

[15] D. Kafetzopoulos, A. Martinou, and V. Bouriotis, *Proc. Natl. Acad. Sci. U.S.A., 90*, 2564 (1993).

[16] S. Tokura, N. Nishi, S. Nishimura, Y. Ikeuchi, I. Azuma, and K. Nishimura, in *Chitin, Chitosan, and Related Enzymes* (J. P. Zikakis, Ed.), Academic Press, Orlando, Florida, 1984, p. 303.

[17] S. Hirano, in Ref. 4, p. 37.

[18] M. G. Peter, J. P. Ley, S. Petersen, M. H. M. G. Schumacher-Wandersleb, and F. Schweikart, in Ref. 12, p. 323.

[19] S. Hirano, in Ref. 4, p. 37.

[20] M. G. Peter, K.-J. Hesse, and K. Mueller, *Chitin, Chitosan und Derivate*, Ministerium für Natur, Umwelt und Landesentwicklung des Landes Schleswig-Holstein, Kiel, 1992.

[21] A. Giannis and T. Kolter, *Angew. Chem., Int. Ed. Engl., 32*, 1244 (1993).

[22] N. Hutadilok, T. Mochimasu, H. Hisamori, K. Hayashi, H. Tachibana, T. Ishii, and S. Hirano, in Ref. 12, p. 289.

OPPORTUNITIES FOR ENVIRONMENTALLY DEGRADABLE POLYMERS

GRAHAM SWIFT

Rohm and Haas Company
Norristown Road, Spring House, Pennsylvania 19477

ABSTRACT

Environmentally degradable polymers, including polymers that are biodegradable, photodegradable, oxidatively degradable, and hydrolytically degradable, have attracted a great deal of attention over the last 20 years or so, particularly as part of plastics waste-management. Undelivered promises, unsatisfied expectations, and unproved claims of degradation have raised serious questions as to the future of this class of polymers. Consequently, we are now at the stage where environmentally degradable polymers need to be carefully evaluated in order to assess what value or benefit, if any, they offer to the polymer industry and the consumer. This assessment must include clear and universally accepted definitions for terms that describe the various environmental degradation pathways, the definition of an acceptable environmentally biodegradable polymer, and methodologies that are able to quantify the degree of degradation. When all these are achieved, it should be possible to recognize and take advantage of the opportunities that are available for environmentally degradable polymers, provided that other requirements such as cost and properties are acceptable. In this article, the key issues of terminology and test methodology are addressed, and, based on the results, some personal prognoses are made on the selective opportunities that may be available for environmentally acceptable degradable polymers.

INTRODUCTION

Synthetic polymers, natural polymers, and modified natural polymers are widely used throughout most of the world today and contribute enormously to the quality of life in the industrial world and are helping to raise living standards in the Third World. Because polymer properties are readily controlled by their chemical composition and manufacturing process, they are used as commodity plastics in such diverse applications as packaging, personal hygiene products, construction of automobiles, computers, houses, clothing, etc.; they are also used as specialty additives in many applications such as water-soluble polymers in detergents, paints, adhesives, concrete, etc., and in medical applications for drug delivery and as temporary and permanent prostheses such as sutures and bone replacements, respectively. Yet, in spite of all these enormous benefits, synthetic specialty polymers and plastics have, in the minds of many people in the general public, legislative bodies, and environmental groups, an overriding connotation of being harmful to the environment, regardless of all their other attributes. The word "environment" is used loosely here to include the human body and the natural environment, since many of the issues encountered by degradable polymers are common to both; this paper, however, is only concerned with the natural environment.

The concern for polymers in the environment is borne out of the association of synthetic plastics and polymers with a waste-management problem that has emerged in the last 20 years. The problem is to some extent real and to some extent a misguided perception, especially with the current commercial and widely used plastic products which were defined and developed for their durability and resistance properties, and, after disposal, they are visible in the environment as litter and they contribute to landfill overcapacity [1]. This problem was acknowledged several years ago by some segments of the polymer industry, and attempts were made to develop environmentally degradable polymers that on disposal, after use, would harmlessly degrade and return to nature. They were anticipated and promoted to be a total solution to the waste-management problem. Unfortunately, no acceptable test methods or definitions were available as guidelines at the time these polymers were introduced, and in many cases they were overpublicized as meeting an admirable objective. The result, when the deficiencies were exposed, was skepticism for the degradable polymer industry and a severe setback for the acceptance of environmentally degradable polymers. Since then, however, with dedicated hard work, industrial and academic scientists have diligently applied themselves to developing meaningful and realistic standard test methods and definitions such that the newer polymers in advanced development and now becoming commercially available should be acceptable in selective applications where environmental degradation is competitive with other recognized waste-management options such as incineration, recycle, and source reduction.

In this paper I will explore opportunities for this new generation of environmentally degradable polymers, the success of which I believe is critically dependent on the general acceptance of the definitions and test methods being developed, and the resolution of environmental safety issues and stigmas associated with the earlier polymers. With these established, only cost and performance, as with any other competitive polymer, will need to be addressed in order to gain market share in appropriate application areas.

ENVIRONMENTALLY ACCEPTABLE DEGRADABLE POLYMERS

Definitions and Test Methodology

Environmentally degradable polymers are generally divided into four different categories which describe their degradation pathways: biodegradation, hydrolytic degradation, photodegradation, and oxidative degradation. They have been defined in many publications and in many different ways. The definitions established by the American Society of Testing and Materials (ASTM) for plastics [2] are probably as close to universal acceptance as any, and these are included here in a more general sense addressing the more general and broader classification of polymers as a starting point.

A *degradable polymer* is designed to undergo a significant change in its chemical structure under specific environmental conditions, resulting in a loss of properties that may vary as measured by standard tests methods appropriate to the polymer and the application in a period of time that determines its classification.

A *biodegradable polymer* is a degradable polymer in which the degradation results from the action of naturally-occurring microorganisms such as bacteria, fungi, and algae.

A *hydrolytically degradable polymer* is a degradable polymer in which the degradation results from hydrolysis.

An *oxidative polymer* is a degradable polymer in which the degradation results from oxidation.

A *photodegradable polymer* is a degradable polymer in which the degradation results from the action of natural daylight.

Unfortunately, the key definition has not yet been addressed: How do we define an environmentally acceptable degradable polymer? This is the issue that must be resolved in order to get acceptance by the general public and other agencies such as regulatory, legislative, environmental groups, etc. The definitions for biodegradable, hydrolytically degradable, oxidatively degradable, and photodegradable polymers refer to a process of degradation, not to a final result. They are all part of a process which leads to establishing the fate and effects of the polymers in the environment, which is the *real issue* of environmental acceptance, regardless of the degradation mechanism. This is sometimes overlooked in the zeal to develop degradable polymers. In this sense, polymers are no different, and should not be treated differently, from any other chemical introduced into the environment. As stated, therefore, none of the accepted definitions for degradation pathways is really of any value beyond the scientist and certainly not to a legislator, an environmentalist, the public, or a developer of new polymers. To develop a definition of an environmentally acceptable degradable polymer, we have to look carefully at the different degradation pathways and determine what questions must be answered in order to ensure that the polymer and its degradation products are safe and pose no harm to the disposal environment.

Figure 1 is a schematic of the four degradation pathways mentioned and what further events beyond initial degradation may occur in the environment. All the degradation mechanisms initially produce fragments which may remain in the environment as recalcitrant pieces with unknown fate and effects or they may be completely biodegraded and ultimately mineralized. The key role of biodegradation is

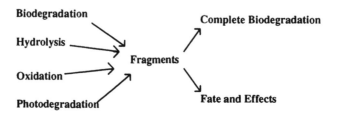

FIG. 1. Pathways for environmentally degradable polymers.

very apparent: it is the only degradation pathway that is able to completely remove a polymer or its degradation products from the environment. Therefore, polymers designed for any of the environmental degradations pathways should also take into consideration that their fragments must either completely biodegrade or be harmless in the environment.

The chemistry of the key degradation process, biodegradation, is represented below by Eqs. (1) and (2), where C represents either a polymer or a fragment from any of the degradation processes defined earlier. For simplicity here, the polymer or fragment is considered to be composed only of carbon, hydrogen, and oxygen; other elements may, of course, be incorporated in the polymer, and these would appear in an oxidized or reduced form after biodegradation depending on whether the conditions are aerobic or anaerobic, respectively.

Aerobic biodegradation:

$$C + O_2 \rightarrow CO_2/H_2O + \text{residue} + \text{biomass} + \text{salts} \tag{1}$$

Anaerobic biodegradation:

$$C \rightarrow CO_2/CH_4/H_2O + \text{residue} + \text{biomass} + \text{salts} \tag{2}$$

Complete biodegradation is when no residue remains, and mineralization is when the original substrate, C in this example, is completely converted into gaseous products and salts. Mineralization is really a very slow process because some of the polymer undergoing biodegradation initially produces biomass, and complete biodegradation and not mineralization is the measurable goal when assessing removal from the environment.

Test methods under development at ASTM [2] and in Japan and Europe for assessing the degree of biodegradation in many potential disposal environments are all based on the above chemical equations. The measurements that are required are the original carbon content of the polymer/fragments and its conversion into gaseous products, residue, and biomass. With these measurements the degree of biodegradation, i.e., the degree of removability of any environmentally degradable polymer from a given test environment, may be assessed. If there are fragments, i.e., incomplete biodegradation, these have to be identified and quantified for assessment of environmental fate and effects, which will be discussed in the next section. Major problems with the current laboratory testing methods make them unsuitable for evaluating polymers that are not rapidly, a few weeks at the most, biodegradable, so-called readily biodegradable. This is unfortunate and, as has been indicated by the author previously [3], requires the development of new test methods for polymers and fragments requiring extended time for biodegradation. These will be the subject of future activities within ASTM, and presumably elsewhere. Inevitably, by their nature, these tests will be expensive and probably will require radiolabeling of polymers with, for example, carbon-14. However, environmental safety demands this price, and the degradable polymer industry will not survive if it ignores these requirements.

There are nonquantitative test practices, developed in ASTM [2], that measure the preliminary degradation of polymers in the mentioned pathways by observing polymer property decay; but since these only lead to fragmentation and not to removal from the environment, they will not be discussed here. Our interest is in what happens to the fragments, and as we have seen they must be tested for fate and effects unless they demonstrably completely biodegrade. The reader interested in test methodology for hydrolysis, oxidation, and photodegradation is referred to the appropriate references for further information.

In order to utilize the information on the biodegradation of polymers and fragments, the test environment must be representative of the disposal environment and the time frame of the disposal method. Since I have already indicated the problem with assessing long-term biodegradation, long-term biodegradable polymers must, at present, be treated as being incompletely biodegradable, and fate and effects in the environment must be established over the period of time in which they are predicted to biodegrade. Hence we must develop new test methods that are predictive of the degree of biodegradation of polymers that are slowly biodegraded so that fate and effects of the environment can be more easily evaluated and an environmental safety assessment (ESA) made. If this cannot be achieved, biodegradable polymers and fragments that degrade over several months and beyond will be difficult to accept for disposal in any environment.

Environmental Safety Assessment

Environmentally degradable polymers and ESAs are related by Eqs. (3) and (4). An ESA on a particular compound is, among other variables, a function of the environmental concentration of the xenobiotic, which in the case of environmentally degradable polymers is related to its concentration and its degradation products at any given time in the degradation cycle.

$$\text{ESA} = f[\text{environmental concentration}] \tag{3}$$

$$\text{ESA} = f[\text{polymer} + \text{degradation products}] \tag{4}$$

The implication, then, is that for any of the degradation pathways, the degradation products must be identified and quantified so that their fate and effects on the environment can be assessed. If the degradation products are completely biodegradable, are identified as naturally occurring or environmentally benign, the polymer is probably acceptable in the environment.

To reemphasize, if fragments remain in the environment after degradation, the polymer is slowly biodegradable, or if the polymer is recalcitrant, it will be necessary to establish the identity and concentrations for all the intermediates and residues so that their fate and effects can be established. This will be an assessment of the no effect concentration (NOEC) of the polymer and/or its degradation products as represented in Eq. (5) by testing or by reference if a particular compound is known. The NOEC is referenced against the most sensitive species that the compounds may be in contact within the environment. The NOEC must be higher than the predicted environmental concentration of the intermediate, and the test method used to establish (predict) the concentration of the compound in the environment is assigned an assessment factor (AF) of from 1 to 1000 depending on its reliability and reproducibility. Real world environmental testing is given a factor of 1, laboratory testing is generally in the range of 10 to 1000. Hence the need for reproducible and reliable test methods and confirmation in real world testing, so that the AF is as low as possible.

$$\text{NOEC/AF} > [\text{environmental concentration}] \tag{5}$$

This requirement for ESAs and the limited testing acceptability of the current test methods beyond a few weeks strongly implies that there is a need to be able to design polymers for waste management that will completely biodegrade (regardless of initial degradation mechanism) within the time frame of the disposal method; for example, a few days in a wastewater treatment facility for water-soluble polymers and a few weeks for composting. In this way the degradation will be complete in the disposal environmental compartment and further testing in subsequent environments will not be necessary. This will save considerable effort that would be necessary for fate and effects testing and environmental safety assessments as well as their associated costs. The concept is captured in the hypothetical degradation profiles shown in Fig. 2. Here the time to completely biodegrade is shown for three different hypothetical polymers which require T1, T2, and T3 time units. The message is that this time to completely biodegrade should be as short as possible and no longer than the time available in the disposal environment. This rate of biodegradation is related to the polymer structure. Once this is understood, it should be possible to modify the rate to meet the desired goal of any particular controlled disposal method.

Considering all these discussed limitations, we are now able to define an environmentally acceptable degradable polymer as:

a degradable polymer that introduces no harmful or toxic residues into the environment either during or after its degradation is complete

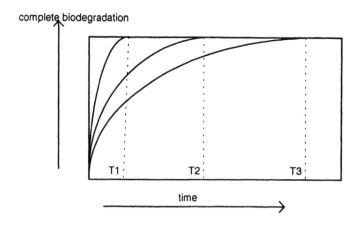

FIG. 2.   Environmental safety assessment and time to biodegrade.

The most expedient way of ensuring this acceptability is for the polymer or fragments from degradation to be demonstrated to be completely biodegradable in the environment in the shortest possible time. Failing to achieve complete biodegradation means identification of residues and fate and effects evaluations which puts a considerable burden of proof on the developer and involves a risk assessment which must meet the scrutiny of many watch committees, including environmentalists, the public, and legislators. Table 1 lists some interesting facts on the public's perception of risk [4] that are worth considering when faced with persuasion of acceptability rather than complete biodegradation. Note that the public will have increased concerns depending on the risk origin, natural or synthetic; trust in authority is lacking; controlled is better than uncontrolled; media attention. They, the public and others opposed to environmentally degradable polymers, are not readily persuaded that a risk is acceptable, and the development of completely removable polymers may be the most prudent course to take with environmentally degradable polymers. This may make the goal much more difficult to achieve, but it will make the opportunities more permanent.

POTENTIAL USES FOR ENVIRONMENTALLY ACCEPTABLE DEGRADABLE POLYMERS

Given the foregoing discussion, we are now in a position to selectively choose potential uses for environmentally acceptable degradable polymers.

It is worth spending a short time to review the history of the use of these polymers in the environment. The early history in this is devoted almost entirely to plastics [5] in waste management. No one, until recently, seemed to recognize that it was a problem shared by all polymers. The first degradable polymers developed were promoted as the panacea for the plastics waste-management problem, promising a rapid and complete solution. Even though the polymers developed as solutions were elegant and very innovative, the environment degradation chemistry at best was only promising, and the polymers that were developed were largely untested in

TABLE 1. Important Factors in the Public's Perception of Risk

Factors	Associated characteristics	
	Increased concern	Decreased concern
Origin	Synthetic	Natural
Familiarity	Unfamiliar	Familiar
Volition	Involuntary	Voluntary
Controllability	Uncontrolled	Controlled
Media attention	Much exposure	Little exposure
History	Yes	No
Trust in authority	Lack of	Trust

the environment with any stringent test methodologies since they were not available. With few exceptions the polymers did not meet the requirements of acceptable environmental degradation, and the whole approach had to be reconsidered.

Currently, the horizon has broadened and environmentally degradable polymers are the focus rather than just plastics, recognizing that all polymers represent potential waste-management problems in the environment. Polymer waste-management is now more distinguished by polymer form, plastics are solids and for the most part readily recoverable after use (litter notwithstanding), and water-soluble polymers are very difficult, if not impossible, to recover after use. The difficulties and misrepresentations of the past have been acknowledged, and definitions and test methods are extant or in development to help in the acceptance process and to avoid unsubstantiated claims; there are no longer claims to a waste-management panacea for these polymers, competition with other options for polymer waste-management is accepted; and selective applications are being sought where the advantages of environmentally degradable polymers make them competitive.

Considering polymer waste-management options as shown schematically in Fig. 3, it is quite apparent that for both types of polymer, plastics and water-solubles, all environmental degradation options are applicable. Plastics, due to their form, are far more readily recoverable and amenable to recycle, either as is or after conversion into monomer or by pyrolysis to new feedstocks for polymers; for incineration as fuel; for disposal by burial; and for composting, which is environmental degradation involving primarily biodegradation, with some contribution by oxidative degradation and hydrolytic degradation. The plastics not recovered are generally considered litter, which may cause land or water pollution with undetermined degradation characteristics and fate and effects in the environment.

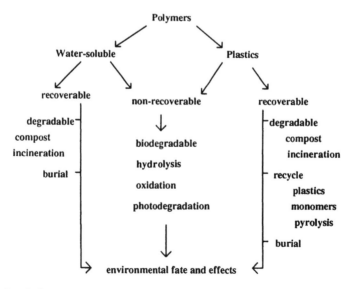

FIG. 3.   Polymer waste-management options.

Water-soluble polymers are generally released into the environment through some type of industrial or municipal wastewater treatment plant. Recovery is only possible by precipitation or by adsorption on the biomass and solids in a treatment facility. Precipitation and recovery is expensive and not likely to be widely practiced, but adsorption on biomass/solids is an option for recovering the polymer for other disposal options such as incineration, burial, environmental degradation in compost, or land application as fertilizer. If not recovered, water-soluble polymers have the option of degrading in the wastewater treatment plant or, if they pass through, the environment beyond. Nondegradables will remain in the environment with unknown fates and effects.

The areas immediately obvious as selective opportunities for environmentally degradable polymers are recoverable plastics and all water-soluble polymers. Plastic litter is a social problem which may be avoided by means other than degradation, however, the plastics that are generally recoverable form a large part of the problem. If these are replaced by environmentally acceptable degradable plastics, the litter problem should be reduced. It should be noted that plastics may utilize all the discussed degradation pathways whereas water-soluble polymers will not usually be susceptible to photodegradation.

In order to realize these opportunities, plastics and water-soluble polymers will have to have properties comparable to the products that they replace, will have to meet the stringent requirements for environmental acceptable degradable polymers, and will have to have some acceptable cost level. This cost factor is currently a major issue since there is no readily apparent agreed upon value for acceptable environmental degradability, and there is every expectation that these specialty polymers, being more chemically sophisticated, will be more expensive than their durable counterparts. Someone has to decide that value; the free market is unlikely to do so. If conscience will not, then legislation will be needed.

One of the major opportunities clearly flagged is that all water-soluble polymers should be biodegradable. Assuming that they all enter the environment through the wastewater treatment facility, the residence time in this compartment is a good target, though difficult, for the complete biodegradation of water-soluble polymers. The residence time will depend on the nature of the polymer. With adsorption on biomass, the polymer has typically 5 to 14 days for degradation to occur, which is the biomass retention time, whereas without adsorption the hydraulic residence time is 3 to 6 hours. Adsorption without complete degradation will allow the opportunity for continued degradation in subsequent environments as mentioned above, such as composting. Specific examples of opportunities are flushable polymers and other personal hygiene products, diapers, superabsorbents, detergent polymers, etc.

Plastics face significant competition from alternative technologies for waste management. The opportunities here are, therefore, more selective. They must be sought in areas where the value added is worth the change to a degradable plastic. Opportunities include packaging for fast-food and other one-way uses where sorting for recycle or incineration is not competitive with composting; agricultural film where labor costs for recycle are high; fishing gear when fish and marine mammal life may be preserved; and personal hygiene products. Litter problems would also benefit from additional plastics that photodegrade and then rapidly biodegrade [6] rather than just photodegrade as with the present plastics.

The significance of all these opportunities is that they must meet the goals of environmentally acceptable degradable polymers discussed here. To do this, there is a need to control the disposal method, the time and mechanism of degradation, and to be cost competitive with the nondegradable polymers currently in use for the targeted applications. This is a significant task. It can be accomplished, but it will take time to satisfy all the requirements for environmentally degradable polymers.

CONCLUSIONS

Environmentally acceptable degradable polymers have been defined as polymers that degrade in the environment by several mechanisms to produce no harmful or toxic residues; it is preferable that the different degradation pathways, biodegradation, hydrolytic degradation, oxidative degradation, and photodegradation, culminate in complete biodegradation so that no residue remains in the environment. This very stringent definition represents a difficult goal for polymer synthesis and is likely to result in degradable polymers having a higher cost than the products they must replace. Nevertheless, a clean, safe environment is the intent, and cost must be calculated by nonclassical methods. Somehow we have to factor in the cost of preserving the environment, something we have only recently begun to consider with commercial products. An example would be the introduction of catalytic converters on automobiles to improve air quality—at significant cost but with consumer support.

If we abide by the definitions discussed here, there are numerous opportunities for acceptable environmentally degradable polymers. These are primarily in the areas where polymers are difficult to recover. Water-soluble polymers are not recoverable economically and are, consequently, a big opportunity. With plastics, niche markets are more likely to be the major opportunities, and these are tied to plastic uses where the products are not readily amenable to alternative waste-management options. Opportunities include fast-food and one-way packaging, and personal hygiene products which are compostable without separation; agricultural films which are mulchable and require no collection; and fishing gear which may break free in the ocean and rivers to cause danger to fish and mammalian life.

REFERENCES

[1]   G. Swift, *Agricultural and Synthetic Polymers*, ACS Symposium Series #433, 1990, p. 2.
[2]   *ASTM Standards on Environmentally Degradable Plastics*, ASTM Publication Code Number (PCN): 1993, #003-420093-19.
[3]   G. Swift, *3rd International Scientific Workshop on Degradable Plastics and Polymers*, Osaka, Japan, November 9–11, 1993.
[4]   V. T. Covello, P. M. Sandman, and P. Slovik, *Communication, Risk Statis-*

tics, and Comparisons: A Manual for Plant Managers, CMA, Washington, D.C., 1988, p. 54.

[5] J. E. Potts, *Aspects of Degradation and Stabilization of Polymers*, Elsevier, 1978, p. 617.

[6] *EPA Standards for Plastic Ring Carriers Environmental Fact Sheet*, EPA 530-F-94-009, February 1994.

USE OF BIOSYNTHETIC, BIODEGRADABLE THERMOPLASTICS AND ELASTOMERS FROM RENEWABLE RESOURCES: THE PROS AND CONS

ALEXANDER STEINBÜCHEL

Institut für Mikrobiologie der Westfälischen Wilhelms-Universität Münster
Corrensstrasse 3, D-48149 Munich, Germany

ABSTRACT

For several reasons, biodegradable polymers have recently attracted much public and industrial interest. This contribution focuses on microbial aspects of the biodegradability of polymeric materials in the environment. It provides an overview on biodegradability and the mechanisms of degradation of relevant natural and technical polymers. Almost all biosynthetic polymers, which are readily available from renewable resources, are biodegradable within a reasonable time scale. Many semibiosynthetic and even chemosynthetic polymers are also biodegradable if they contain chemical bonds which also occur in natural compounds. Only a few polymers are truly persistent. Arguments for and against the use of such polymers are summarized.

INTRODUCTION

Polymers represent the most abundant class of organic molecules of our biosphere [1]. Complex polymers like coal and lignin occur in amounts of approximately $835,000 \times 10^6$ and $700,000 \times 10^6$ tonnes, respectively. These polymers represent the largest fractions, and they were formed in the past or synthesis is still going on, respectively. Furthermore, approximately $40,000 \times 10^6$ tonnes cellulose and $20,000 \times 10^6$ tonnes lignin are currently synthesized each year. In contrast to these figures, the production of approximately 100×10^6 tonnes per annum of various plastic materials is rather low. However, although only comparably small amounts of plastics are produced, these materials are the source of rising problems since most of them are not used for long-lived materials but occur in waste, and communities in many countries have difficulty managing the problems resulting from continuously increasing amounts of plastic wastes.

Biodegradable polymers have recently attracted much public and industrial interest as a consequence of extensive discussions looking for better waste-management strategies. The use of biodegradable polymers allows composting as an additional way for waste disposal, which is currently done mostly by deposition in landfills or by incineration. Furthermore the use of biodegradable polymers opens several new applications in medicine, agriculture, and many other areas. Therefore, scientists and industry are intensively looking for polymers which are biodegradable in their respective habitats. On the other hand, biodegradability of a polymer may limit its use for long-lived materials, since the latter should not only be resistant to chemical degradation and physical disintegration but also to biodegradation.

CLASSIFICATION OF POLYMERS

Because of their chemical structure, polymers represent a rather complex class of molecules. Besides linear, branched, and crosslinked polymers, homopolymers, various kinds of copolymers (random, alternating, block, and grafted), as well as blends of various types can be distinguished (Fig. 1).

From their origin we can distinguish completely biosynthetic and completely chemosynthetic polymers (Table 1). In addition, many semibiosynthetic polymers are known; these polymers are obtained from chemical polymerization of precursors of which at least one results from biosynthesis. Also, many semichemosynthetic

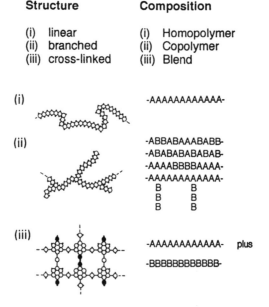

FIG. 1.

TABLE 1.   Classification of Polymers (ii)

Completely biosynthetic   [examples: proteins, polysaccharides, starch plastics, PHA]

Semibiosynthetic: Biosynthesis of at least one precursor and subsequent chemical polymerization [examples: polylactides, polyacrylamide, polyurethane, Bionolle (potentially), polymeric derivatives of the degradation of aromatic compounds]

Semichemosynthetic: Chemosynthesis of precursors and subsequent biological polymerization [example: PHF]

Biosynthetic but chemically modified   [examples: casein plastics, cellulose nitrate, crosslinked PHA]

Chemosynthetic but biologically modified   [?]

Completely chemosynthetic   [examples: polyethylene, polyurethane, Bionolle]

---

polymers are known; these polymers are synthesized biologically from precursors which result from chemical synthesis. Furthermore, biosynthetic polymers which have been chemically modified upon biosynthesis are known.

BIODEGRADATION

Many organisms are able to degrade polymers which occur outside the cells. Among polymer degrading organisms, bacteria and fungi are the most abundant and exhibit the most flexible metabolism. Degradation of polymers—especially if plastic is degraded—often requires the disintegration of supramolecular structures (Step 1). Only after this first step will the backbone of the polymer become susceptible for cleavage by enzymes (Step 2). This step occurs mostly outside the cell by enzymes which are excreted into the medium. It has not yet been conclusively shown that polymers are taken into the cell prior to cleavage. Subsequently, monomeric or oligomeric cleavage products are transported into the cell (Step 3) and are degraded by catabolic enzymes (Step 4). With aerobic organisms, carbon dioxide and water may be the only degradation products of the polymer. With anaerobic organisms, fermentation products like organic acids and alcohols as well as methane may occur in addition to carbon dioxide [2].

Only the subsequent combination of all four steps will result in complete degradation of the polymer. If only Step 1 has occurred, the polymer may become invisible to the naked eye although the polymer molecules still exist and although degradation of the polymer has not occurred. This type of "biodegradation" often occurs with materials made from a blend of a persistent polymer and a biodegradable polymer; e.g., a blend of polyethylene and starch, respectively.

NATURALLY OCCURRING MECHANISMS FOR
CLEAVAGE OF POLYMERS

Polymers represent major constituents of the living cells which are most important for the metabolism (enzyme proteins, storage compounds), the genetic information (nucleic acids), and the structure (cell wall constituents, proteins) of cells [2]. These polymers have to be degraded inside cells in order to be available for environmental changes and to other organisms upon cell lysis. It is therefore not surprising that organisms, during billions of years of evolution, have developed various mechanisms to degrade naturally occurring polymers. It is therefore also not surprising that all biosynthetic polymers, like all other natural products, are more or less readily degraded by organisms [3]. During the last hundred years the chemical industry has added many new polymers which are used as thermoplastics, elastomers, duroplasts, or many other applications. Most of these polymers are not, in general, persistent to biodegradation [4-6]. Only a few chemosynthetic polymers (e.g., polyethylene) are completely resistant to microbial degradation [7].

Since degradation of polymers occurs mostly outside cells, the number of mechanisms available for cleavage of the polymer backbone is limited ([6], Table 2). Glycosidic bonds as well as peptide bonds and most ester bonds (e.g., in proteins, nucleic acids, polysaccharides, and polyhydroxyalkanoic acids) are cleaved by hydrolysis. Some glycosidic bonds in polysaccharides are cleaved by a phosphorolytic mechanism; the enzymes cellobiose phosphorylase and α-1,4-glucanphosphorylase for the phosphorolytic cleavage of cellulose or starch, respectively, are examples. Lignin peroxidase catalyzes the cleavage of carbon—carbon bonds and of ether bonds in lignin. The enzyme has been studied in most detail in the fungus *Phanerochaete chrysosporium* [8]. Its unique radical mechanism seems to allow for attack on other molecules which are not readily degraded or which are not degraded at all. An oxidative mechanism catalyzed by polyethylene glycol dehydrogenase of various aerobic bacteria, which obviously depends on the presence of the artificial electron acceptor dichlorophenol indophenol, and a mechanism catalyzed by polyethylene glycol acetaldehyde lyase of various anaerobic bacteria employing rearrangements of carbon—carbon bonds and depending on cobalamin, were recently described for the degradation of water-soluble polyethylene glycols. From theoretical considerations and from experimental studies, evidence for an intracellular degradation of polyethylene glycol was obtained. How the polymer molecules pass the cell wall is, however, not known [6].

ARGUMENTS FOR THE USE OF BIOSYNTHETIC, BIODEGRADABLE POLYMERS FROM RENEWABLE RESOURCES

The use of thermoplastics and elastomers which are biotechnologically produced from renewable resources may offer several advantages as compared to the

TABLE 2. Mechanisms for Enzymatic Cleavage of the Backbones of Polymers

Mechanism	Cleavage of	Enzyme
Hydrolytic	Glycosidic bonds, ester bond, peptide bonds	e.g., proteases, nucleases, cellulases, muramidases
Phosphorolytic	Glycosidic bonds	e.g., cellobiose-phosphorylase
Radical	Carbon—carbon bonds, ether bonds	Lignin-peroxidase
Oxidative	Ether bonds	Monooxygenases, PEG-DH
Rearrangements	Carbon—carbon bonds	PEG-acetaldehyde lyase

use of conventional polymers. Production of these polymers is independent of fossil resources; fossil carbon sources are only required to support the growth (e.g., for the production of fertilizers and herbicides), and the harvest of plants provides the carbon sources that are offered to the bacteria in the bioreactor and the energy for the manufacture of materials from the polymer by, e.g., extrusion [9]. The carbon atoms of the polymer are derived from carbon dioxide in the atmosphere, and the synthesis of the renewable resource is driven by solar energy. The use of renewable resources will provide an alternative to the production of food for agriculture. At present, farmers in most developed countries produce agricultural surpluses. In addition, many regions are no longer suitable for the production of food since they were heavily polluted in the past. Land which is no longer required and which is not suitable for the production of food may therefore be used for the production of renewable resources. In addition, many interesting polymers could be produced from waste materials derived from food industry processes or from the paper manufacturing industry. In our laboratory, for example, recombinant strains of *Alcaligenes eutrophus* were obtained which utilize lactose (abundant in whey) as a carbon source for the production of polyhydroxylalkanoic acids [10]. Many more examples could be provided. Finally, and as outlined below, the use of some thermoplastics and elastomers may lead to many new applications.

One major argument for the use of these polymers is that waste can be directed to composting. Materials manufactured from these polymers must no longer be deposited in landfills, which are increasingly less available. In contrast to incineration, chemical reutilization, or recycling, composting can be achieved at comparably low capital costs, and it does not require extensive transportation. On the other hand, biodegradable materials disappear rather than being litter.

ARGUMENTS AGAINST THE USE OF BIOSYNTHETIC, BIODEGRADABLE POLYMERS FROM RENEWABLE RESOURCES

Despite the above-mentioned advantages, the use of biosynthetic, biodegradable polymers and their production from renewable resources is very controversial. The primary reason why biosynthetic polymers from renewable resources are not as widely used as chemosynthetically polymers produced from fossil resources is the high costs of these polymers. The price of Biopol, which is the trade name for a bacterial copolyester consisting of 3-hydroxybutyrate and 3-hydroxyvalerate, is approximately 15 to 20 US dollars per kilogram. This is much higher than the cost for conventional plastics used for the manufacture of packaging material. However, the price for Biopol will most likely drop to approximately 4 to 8 US dollars in large-scale production [11]. The situation is similar for other biosynthetic polymers.

Other concerns arise from the lack of evidence for a positive "Ökobilanz." It is doubted that the total input of energy is less and that all effects on the biosphere are less severe than for the production of polymers from petrochemicals. These complex effects have to be studied in detail. One has, however, to consider that the processes for the production of polymers from renewable resources may change drastically as a consequence of the increase of knowledge on the biosynthetic pathways and on the production organisms, and also due to the application of recombinant DNA technology. Many bacteria are able to synthesize large amounts of polyesters consisting of a wide range of hydroxyalkanoic acids [12, 13]. These polyesters have thermoplastic and elastomeric properties and are biodegradable. The polyhydroxyalkanoic acids biosynthesis genes from various bacteria have recently been cloned and analyzed at the molecular level [14]. For example, it was shown that it is possible to produce at least some polyhydroxyalkanoic acids in transgenic plants utilizing carbon dioxide and solar energy directly [15].

One other major and general concern on the use of biodegradable polymers is the possibility that greenhouse gases other than carbon dioxide are released into the atmosphere during the production and after the use of biodegradable polymers. Support of plant growth for the production of renewable resources by fertilizers may result in an increased release of NO_x gases. If biodegradable materials are deposited in landfills, the polymers will be at least partially converted to methane in anoxygenic areas by consortia of quite different organisms including methanogenic bacteria [2]. Methane, for example, exerts an approximately 20-fold higher greenhouse effect than does carbon dioxide.

The use of biodegradable polymers for recently developed plastics is further limited by the poor physical and technical properties of these materials. This is at present certainly true for many polymers and applications. However, one has to consider that such products as Biopol, Bionolle, and starch plastics were developed only recently, whereas most conventional plastics in wide use were developed over a period of several decades and were continuously improved. This will probably also occur in the future with the materials mentioned above. For example, the barrier properties of films manufactured from biodegradable polymers toward water vapor and molecular oxygen can be improved by metal vapor treatment as shown for PHA films treated with Al or SiO_x [16].

Certainly the use of biodegradable polymers will remain restricted to applications for short-lived applications. These polymers will probably not be useful for the outside parts of cars, for the construction of buildings, for the manufacture of pipelines, etc. Furthermore, the use of biodegradable materials for food packaging might cause hygienic problems under certain circumstances if the necessary precautions are not taken.

NEW APPLICATIONS FOR BIODEGRADABLE POLYMERS

Other applications of biodegradable polymers in the packaging industry focus on their use to improve the resistance to water and humidity of other materials. Plasticized starch has the disadvantage of not being waterproof. The use of blends of starch and water-insoluble PHA may overcome this problem. Water-sensitive materials can be also coated with PHA [17], allowing the replacement of barrier layers of aluminum or polyethylene in cardboard packages for soft drinks and mil, drinking cups, etc.

The uses of biodegradable polymers are not restricted to applications in the packaging industry which require large amounts of polymers produced at reasonable low costs. There are many other potential applications in other areas which require less material and for which the distinct properties of these polymers rather than their costs are important. Some of these applications will, of course, remain niche applications. For example, polyhydroxyalkanoic acids may be considered for many interesting applications in medicine, agriculture, and the food industry or to provide a source of interesting substrates for the chemical synthesis of enantiomeric pure chemicals [18]. Even applications for polyhydroxyalkanoic acids in photoelectrical apparatuses are possible. For example, nonbiodegradable thermoplasts are currently used in xerography where they function as carriers for the toner and cover a significant fraction of the surface of the principally biodegradable paper. This does not contribute to the biodegradability of paper during composting. Since the toner is tightly fixed to the paper surface, de-inking of the paper becomes a problem during recycling of used paper. It is currently being investigated whether biodegradable polyhydroxyalkanoic acids can be used instead of conventional nonbiodegradable thermoplastics.

CONCLUSIONS

The field of biosynthetic biodegradable polymeric materials has been largely developed in recent years, and many new polymers have been made available. Considering PHA, only poly(3-hydroxybutyrate) was known for almost six decades. Systematic feeding of precursor substrates to various bacteria, cultivation of the bacteria under appropriate conditions, and improved methods for the analysis of the accumulated polyesters have revealed approximately 80 different constituent hydroxyalkanoic acids of PHA. Most of these new PHAs have been detected in *Alcaligenes eutrophus* and *Pseudomonas oleovorans* only during the last 5 years [12, 13]. The PHA syntheses of both bacteria are distinguished by rather low substrate specificities. Furthermore, heterologous expression of PHA syntheses genes in other bacteria, which provide a suitable physiological background, resulted in additional new polymers [19–21]. In addition, biosynthetic polyhydroxyalkanoic acids can be modified by various measures after biosynthesis has been completed and after the polyesters have been isolated from cells. If polyhydroxyalkanoic acids consisting of medium-chain-length hydroxyalkanoic acids are exposed to irradiation, a rubberlike complex polymer is obtained which is still biodegradable [22]. The evaluation of the physical, technical, and biological properties of biosynthetic PHA is still at the very beginning, and analysis of the data will probably give hints for many new applications.

REFERENCES

[1] M. Eggersdorfer, S. Warwel, and G. Wulff, *Nachwachsende Rohstoffe. Perspektiven für die Chemie*, VCH Verlagsgesellschaft, Weinheim, 1993.

[2] H. G. Schlegel, *General Microbiology*, Cambridge University Press, London, 1993.

[3] G. Winkelmann, *Microbial Degradation of Natural Products*, VCH Verlagsgesellschaft, Weinheim, 1992.

[4] R. B. Cain, *Microbial Control of Pollution* (J. C. Fry, G. M. Gadd, R. A. Herbert, C. W. Jones, and I. A. Watson-Craik, Eds.), Cambridge University Press, London, 1992, pp. 293–338.

[5] M. R. Timmins and R. W. Lenz, *Trends Polym. Sci., 2*, 15–19 (1994).

[6]   B. Schink, P. H. Janssen, and J. Frings, *FEMS Microbiol. Rev., 103*, 311–316 (1992).

[7]   A. C. Albertsson, *J. Macromol. Sci. – Pure Appl. Chem., A30*, 757–765 (1993).

[8]   E. Odier and I. Artraud, in *Microbial Degradation of Natural Products* (G. Winkelmann, Ed.), VCH Verlagsgesellschaft, Weinheim, 1992, pp. 161–192.

[9]   H. G. Schlegel, *FEMS Microbiol. Rev., 103*, 347–354 (1992).

[10] A. Pries, A. Steinbüchel, and H. G. Schlegel, *Appl. Microbiol. Biotechnol., 33*, 410–417 (1990).

[11] D. Byrom, in *Biomaterial, Novel Materials from Biological Sources* (Stockton Press), 1991, pp. 333–360.

[12] A. Steinbüchel, in *Biomaterials* (D. Byrom, Ed.), Macmillan, New York, pp. 123–213.

[13] A. Steinbüchel, *Nachr. Chem. Tech. Lab., 39*, 1112–1124 (1991).

[14] A. Steinbühcel, E. Hustede, M. Liebergesell, A. Timm, U. Pieper, and H. Valentin, *FEMS Microbiol. Rev., 103*, 217–230 (1992).

[15] Y. Poirier, D. Dennis, K. Klomparens, C. Nawrath, and C. Somerville, *Ibid., 103*, 237–246 (1992).

[16] C. Bichler, R. Drittler, H. C. Langowski, A. Starnecker, and H. Utz, *BioEngineering, 9*, 9–16 (1993).

[17] R. H. Marchessault, P. Rioux, and I. Saracovan, *Nord. Pulp Pap. Res. J., 1*, 211–216 (1993).

[18] A. Steinbüchel, *Chem. Labor Biotech., 44*, 378–384 (1993).

[19] M. Liebergesell, F. Mayer, and A. Steinbüchel, *Appl. Microbiol. Biotechnol., 40*, 292–300 (1993).

[20] H. E. Valentin and A. Steinbüchel, *Ibid., 40*, 699–709 (1994).

[21] H. E. Valentin, E. Y. Lee, C. Y. Choi, and A. Steinbüchel, *Ibid., 40*, 710–716 (1994).

[22] G. J. M. de Koning, H. M. M. van Bilsen, P. J. Lemstra, W. Hazenberg, B. Witholt, J. G. van der Galien, A. Schirmer, and D. Jendrossek, *Polymer, 35*, 2090–2097 (1994).

BIODEGRADATION OF POLYMERS AT TEMPERATURES UP TO 130°C

G. ANTRANIKIAN, A. RÜDIGER, F. CANGANELLA, M. KLINGEBERG, and A. SUNNA

Department of Technical Microbiology
Institute of Biotechnology
Technical University Hamburg-Harburg
Denickestrasse 15, 21071 Hamburg, Germany

ABSTRACT

Extreme thermophilic and hyperthermophilic microorganisms are those which are adapted to grow at temperatures from 70 to 110°C. Most of these exotic microorganisms are heterotrophic and are capable of attacking various polymeric substrates such as starch, hemicellulose, and proteins. Only recently, a number of novel extracellular enzymes like α-amylase, pullulanase, xylanase, and proteinase have been purified and studied in detail. By applying gene technology it was also possible to purify heat-stable enzymes after expression of their genes in mesophilic hosts. These novel enzymes are in general characterized by temperature optima around 90–105°C and a high degree of thermostability. Enzymic activity is still detectable even at 130°C and in the presence of detergents. Due to the remarkable properties of these enzymes, they are also of interest for biotechnological applications.

INTRODUCTION

A number of thermophilic microorganisms were found to produce extracellular enzymes that are capable of hydrolyzing polymers such as starch, cellulose, hemicellulose, and proteins. The majority of these organisms grow optimally between 60 and 75°C. Little, however, is known on the enzymology of hyperthermo-

69

philic microorganisms which grow optimally above 85°C. Only in the last decade has it been possible to isolate microorganisms which can grow optimally even above 100°C [1, 2]. The majority of this group that thrive above the boiling temperature of water belong to the archaea. However, some of these microorganisms also belong to the bacterial kingdom. The thermophilic representatives of the bacteria that optimally live above 65°C comprise four genera, namely, *Thermotoga*, *Thermosipho*, *Fervidobacterium* (Thermotogales order) and *Aquifex* (Aquificales order). The temperature optimum for growth of these microorganisms ranges between 65 and 90°C. On the other hand, the thermophilic representatives of the archaea comprise more than 19 genera which belong to the following orders: Sulfolobales, Pyrodictiales, Thermoproteales, Thermococcales, Archaeglobales, Thermoplasmales, and the methanogens Methanobacteriales and Methanococcales. The majority of the microorganisms are heterotrophic and anaerobic; only a few are strict autotrophes (Table 1). Most of these exotic microorganisms have been isolated from various geothermal habitats like hot springs, sulfataric fields, and deep-sea hydrothermal vents. Of great interest are the enzymes that are formed by extreme thermophilic and hyperthermophilic microorganisms. Some of the enzymes that have been recently studied are even active at 130°C [3]. Polymeric substrates are abundant in nature and provide a valuable and renewable source of carbon as well as of energy for various microorganisms that sustain life at high temperatures. These organisms are capable of attacking complex polymeric substrates by producing enzymes with a wide range of specificity.

STARCH DEGRADATION

Starch is composed of amylose (15–25%) and amylopectin (75–85%). Amylose is a linear macromolecule consisting of 1,4-linked α-D-glucopyranose residues. The chain length varies from several hundred to 6000 residues. The direction of the chain is characterized by the reducing and the nonreducing end. The reducing end is formed by a free C1 hydroxyl group. Like amylose, amylopectin is composed of α-1,4-linked glucose molecules, but in addition branching points with α-1,6-linkages occur. The branching points occur at every 17–26 glucose molecules, so that the content of α-1,6-linkages in amylopectin is about 5%. Due to the molecular mass of 10^6 to 10^9, amylopectin is one of the largest biological molecules.

The enzymes involved in the conversion of this substrate to low molecular weight compounds like glucose, maltose, and oligosaccharides with various chains are α-amylase, β-amylase, glucoamylase, debranching enzymes (pullulanases), and α-glucosidase. Enzymes that are capable of hydrolyzing α-1,6-glycosidic linkages in pullulan and amylopectin are defined as debranching enzymes or pullulanases. The coordinate action of many enzymes is usually needed for the efficient conversion of macromolecules.

The capability to utilize starch as carbon and energy source is widely distributed among microorganisms like bacteria, fungi, and yeasts [4]. It has been also shown that starch stimulates growth of a number of extreme thermophilic and hyperthermophilic microorganisms. Very recently a number of extreme thermophilic and hyperthermophilic microorganisms were found to produce novel thermoactive amylolytic enzymes with peculiar properties. These microorganisms belong to

TABLE 1.   Taxonomy and Some Biochemical Features of Bacteria and Archaea Growing at High Temperatures[a]

Order	Genus	Optimal growth temperature, °C	Heterotrophic (het) Autotrophic (aut) Facultative autotrophic (f)	Anaerobic (an) Aerobic (ae)
		Bacteria		
Thermotogales	*Thermotoga*	70–80	het	an
	Thermosipho	70–75	het	an
	Fervidobacterium	65–70	het	an
Aquifecales	*Aquifex*	90	het	ae/an
		Archaea		
Sulfolobales	*Sulfolobus*	65–80	f	ae/an
	Metallosphaera	75	f	ae
	Acidianus	88	aut	ae/an
	Desulfurolobus	80	het	ae/an
Pyrodictales	*Pyrodictium*	100–105	het, aut	an
	Thermodiscus	88	f	an
	Hyperthermus	100	het	an
Thermoproteales	*Thermoproteus*	88	het, f, aut	an
	Thermofilum	88	het	an
	Desulfurococcus	85	het	an
	Staphylothermus	92	het	an
	Pyrobaculum	100	het, f	ae, an
Thermococcales	*Thermococcus*	70–87	het	an
	Pyrococcus	100	het	an
Archaeoglobales	*Archaeoglobus*	83	f	an
Thermoplasmales	*Thermoplasma*	60	het	ae/an

[a]Methanogenic microorganisms (Methanobacteriales and Methanococcales) with thermophilic representatives are not shown.

the genera *Thermotoga, Desulfurococcus, Staphylothermus, Thermococcus, Pyrococcus,* and *Sulfolobus* [3]. Detailed studies could be performed after purification of a number of these enzymes to homogeneity. The most thermostable amylase known so far has been purified from *Pyrococcus woesei* and *P. furiosus* [3, 5]. *P. woesei* and *P. furiosus*, which grow optimally on starch at 100°C, possess α-amylase, pullulanase, and α-glucosidase [5, 6].

In order to purify α-amylase from *P. woesei*, cell-free supernatant was concentrated and the enzyme was adsorbed onto soluble starch at 4°C. Release of the enzyme from starch was achieved by boiling and final desorption by electroelution. The native enzyme is composed of a single polypeptide chain with a molecular mass of 68 kDa. Enzymic activity is detected even after treatment of the purified enzyme with SDS (1%) and mercaptoethanol (2%) for 10 minutes at 100°C. This enzyme is capable of hydrolyzing randomly the α-1,4-glycosidic linkages in various glucose polymers such as amylopectin, glycogen, and amylose, forming various oligosaccharides. In addition to soluble polysaccharides, the α-amylase of *P. woesei* can hydrolyze native starch efficiently. Unlike the enzyme from other sources, glucose is not formed as an end product. The smallest substrate that can be attacked by the enzyme is maltoheptaose (DP7) which is converted to DP2 and DP5. Since this organism is unable to utilize glucose as the carbon source, it is highly likely that the products of starch degradation (α-limit dextrins) are directly transported into the cell and further hydrolyzed intracellularly. This novel enzyme displays a temperature optimum of 100°C and is active from 40 to 140°C. Almost 20% of α-amylase activity was detected even at 130°C. For the complete inactivation of the α-amylase, 10 hours of autoclaving at 120°C is necessary. The pH optimum for the enzymatic activity is 5.5; around 50% of activity is measured at pH 4.5 and 7.0. From these data it is evident that the prevailing growth conditions in geothermal habitats are optimal for this extracellular enzyme released into the environment. Additional in-vitro studies have shown that the addition of metal ions is not required for the catalytic activity of the purified enzyme.

The addition of 1 to 5 mM of Cr^{2+}, Cu^{2+}, Fe^{2+}, and Zn^{2+} cause enzyme inhibition; Ca^{2+} ions (up to 7 mM), however, cause slight stabilization of the enzyme. Analysis of the amino acid composition of α-amylase shows no unusual features when compared with the enzymes from mesophilic and thermophilic bacteria.

Another strategy for the purification of heat stable enzymes is the application of gene technology. Attempts were made to clone genes, encoding thermostable enzymes in a mesophilic host. Recently, the DNA encoding of the extremely thermostable pullulanase from *P. woesei* was cloned and expressed in *Escherichia coli*. Due to the thermostability of the pullulanase, it was possible to purify the enzyme in a most remarkable way. Purification of the expressed enzyme was achieved by boiling the fermentation broth, denaturing of the host proteins, and recovery of the thermostable enzyme in the supernatant (Table 2). The activity of the enzyme is highest at 100°C, and 40% of enzymatic activity is detected at 120°C; the pH optimum is 5.5 (Fig. 1). No difference was observed in the physiochemical properties of the native and the cloned enzyme. Unlike the α-amylase from the same organism, the pullulanase activity was stimulated by the addition of calcium ions. The addition of 0.2 mM of Ca^{2+} causes an increase in the enzyme activity of up to fourfold. The pullulanase of *P. woesei* is capable of hydrolyzing α-1,6-linkages and α-1,4-linkages

TABLE 2. Purification of the Cloned Pullulanase from *Pyrococcus woesei*

Step	Total protein, mg	Total activity, U	Specific activity, U/mg	Recovery, %
Cell-free extract	1530	230	0.15	100
Heat treatment	240	180	0.75	79
Maltotriose-Sepharose	9.2	80	8.6	34
Mono Q	2.5	34	13.5	15

in branched oligo- and polysaccharides and hence can be classified as pulluanase type II (also named amylopullulanase). The hydrolysis products are glucose, maltose, and various linear oligosaccharides. HPLC analysis also shows that the pullulanase is able to attack α-1,6-linkages of pullulan in an *endo* fashion, forming a mixture of DP3, DP6, DP9 and DP12 (DP is degree of polymerization). Unlike pullulanases known so far, the thermoactive enzyme from *P. woesei* is also able to attack the α-1,6-linkage in panose, forming maltose and glucose as final products. Due to the multiple specificity of this enzyme, its action causes the complete and efficient conversion of starch to small sugars without the requirement for other amylolytic enzymes. Since the above-mentioned hydrolytic enzymes are optimally active under the same condition, they can be applied in a one-step process for the industrial bioconversion of starch. The improvement of the starch conversion process by finding new efficient and thermoactive enzymes would therefore significantly lower the cost of sugar syrup production.

XYLAN DEGRADATION

Hemicelluloses are noncellulosic low molecular weight polysaccharides that are found together with cellulose in plant tissues. In the cell walls of land plants,

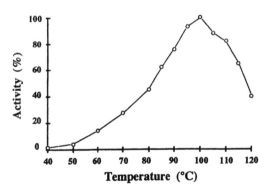

FIG. 1. Influence of temperature on the activity of the cloned pullulanase from *Pyrococcus woesei*; incubation time was 30 minutes.

xylan is the most common hemicellulosic polysaccharide, representing more than 30% of the dry weight. Most xylans are heteropolysaccharides which are composed of 1,4-linked β-D-xylopyranosyl residues. This backbone chain is substituted with acetyl, arabinosyl, and glucuronosyl residues. Due to the heterogeneity of xylan, its hydrolysis requires the action of a xylanolytic enzyme system which is composed of β-1,4-endoxylanase, β-xylosidase, α-L-arabinofuranosidase, α-glucurónidase, and acetyl xylan esterase activities. The concerted action of these enzymes converts xylan to its constituent sugars. Xylan degrading enzymes have been reported to be present in marine and terrestrial bacteria, rumen and ruminant bacteria, fungi, marine algae, protozoa, snails, crustaceans, insects, and seeds of terrestrial plants.

Thermoactive xylanases, arabinofuranosidase, and β-xylosidase have been recently characterized from extreme thermophilic bacteria, namely from *Thermotoga maritima*, *T. neapolitana*, *T. thermarum*, and *Thermotoga sp.* FjSS3-B.1 [11; Sunna et al., submitted]. Xylanase which was isolated from the latter strain has a molecular mass of 31 kDa and is optimally active at pH 5.0–5.5 and 100°C. The xylanase of *T. neapolitana*, *T. thermarum*, and *T. maritima* and β-xylosidase from *T. thermarum* show optimal activity at 90°C. Above 80% of enzymatic activity is still detectable at 105°C. The arabinofuranosidase from the latter organism is less thermoactive and shows maximal activity at 70°C. Substrate specificity tests with the enzyme system of *T. thermarum* indicated that the commercial birchwood xylan as well as xylan from an industrial process (Lenzing AG) were the most suitable substrates. Oat spelt and larchwood xylan were less suitable. When incubated with hydroxyethyl cellulose or filter paper, the xylanolytic enzyme system did not show cellulase activity. The xylanase from *T. thermarum* is capable of hydrolyzing insoluble beech xylan as well as partially soluble birchwood xylan to yield mainly xylobiose, xylotriose, xylotetraose, and longer chains of xylooligosaccharides. This therefore proves the presence of a depolymerizing endo-1,4-β-xylanase. The molecular mass of the major protein band with xylanolytic activity is 40 kDa. An attractive application of such enzymes would be enzyme-assisted pulp bleaching.

PROTEIN DEGRADATION

Proteins are the most abundant organic molecules in living cells and constitute more than 50% of their dry weight. The molecular weight of proteins that are made up of one or more polypeptide chains can vary from a few thousands to more than one million. The three-dimensional conformation of proteins may vary. Globular proteins (spherical or globular) are soluble and usually have dynamic function. Fibrous proteins, on the other hand, occur as sheets or rods, are insoluble, and serve as structural elements. The enzymes which hydrolyze the peptide bonds in proteins are defined as proteinases. They are classified into four groups depending on the nature of their active center.

I. Serine proteinases have a serine residue in their active center and are inhibited by DFP (diisopropylphosphofluoride) and PMSF (phenylmethylsulfonylfluoride).

II.   Cysteine proteinases have a SH groups in their active center and are inhibited by thiol reagents, heavy metal ions, alkylating agents, and oxidizing agents.
III. The activity of metal proteinases depends on tightly bound divalent cations. They are inactivated by chelating agents.
IV.   Aspartic proteinases (acid proteinases) are rare in bacteria and contain one or more aspartic acid residues in their active center.

Inactivation of the enzyme can be achieved by alkylation of the aspartic acid residues with DAN (diazoacetyl-DL-norleucine methyl ester) [7].

Several extreme thermophilic and hyperthermophilic microorganisms which were isolated from submarine hydrothermal vents, solfataric fields, and alkaline springs grow preferentially on complex media containing proteins and peptides, and hence produce thermoactive proteinases. Thermoactive proteinases have been identified very recently from a number of extreme thermophilic and hyperthermophilic archaea like *Pyrococcus furiosus, P. woesei, Desulfurococcus mucosus, Thermococcus celer, T. stetteri, T. litoralis, Thermococcus* sp. AN1, *Staphylothermus marinus, Sulfolobus acidocaldarius*, and *S. solfataricus*. Most of the enzymes formed seem to be associated to the cells. From *P. furiosus* and *P. woesei* a serine-type proteinase (pyrolysin) has been identified and characterized [8–10]. Pyrolysin is a cell-envelope associated proteinase and displays high activity at 110°C; the half-life at 100°C is 4 hours. The enzyme system is active between pH 6.5 and 10.5. Zymogram staining shows the presence of multiple protease bands ranging from 65 to 140 kDa. Further experiments with cell extracts of *P. furiosus* have shown that at least two proteinases are resistant to SDS denaturation. Enzyme resistance to detergents and the appearance of multiple bands in polyacrylamide gels seems not to be unusual for thermoactive enzymes. Other serine proteases were detected in the archaea *T. celer, T. stetteri, Thermococcus* sp. AN1, *T. litoralis*, and *S. marinus* [12]. These enzymes demonstrate optimal activities between 80 and 95°C. The pH optimum is either neutral (*T. celer, T.* AN1) or alkaline. These proteinases exhibited preference to phenyl alanine in the carboxylic site of the peptide. Zymogram staining also shows multiple bands for all strains investigated.

Few proteinases from hyperthermophiles, however, have been purified and studied in detail. The enzyme (archaelysin) which was purified from *Desulfurococcus mucosus* has a molecular mass of 52 kDa and is optimally active at 100°C [13]. Substrate specificity studies of this serine-type enzyme suggest its preference for hydrolytic residues on the C-terminal side of the splitting point. Only recently a serine-type proteinase from *T. stetteri* was purified by preparative SDS-gel electrophoresis. The proteinase, with an apparent molecular mass of 68 kDa, was purified 67-fold. Maximal activity is measured at 85°C, and 40% of activity is measurable at 100°C. The enzyme is active at a broad pH range between 5 and 11. For mapping the P1 binding site, different N-protected p-nitroanilides were tested. The derivatives of glycine, alanine and aspartic acid were not hydrolyzed. The highest activity was obtained with the derivatives of arginine and phenylalanine. The determined kinetic constants are presented in Table 3; the enzyme has esterase activity. The K_{cat}/K_m value for Z-Phe-ONp is four orders of magnitude higher when compared to Suc-Phe-pNA. The amino acid anilides with nonprotected amino groups are not hydrolyzed. Interestingly, the enzyme at various temperatures is highly stereoselective and it hydrolyzes exclusively the L-forms of Bz-Arg-pNa, phenyl glycine amide,

TABLE 3.   Kinetic Parameters for the Hydrolysis of Amino
Acid Derivatives at 80°C, Catalyzed by the Purified Protease
from *Thermococcus stetteri*[a]

Substrate, mM[b]	K_m, mM	k_{cat}, s^{-1}	k_{cat}/K_m, M$^{-1} \cdot$ s^{-1}
Suc-L-Phe-pNA	0.9	0.1	110
H-L-Phe-pNA	0	0	0
Bz-L-Arg-pNA	0.06	0.013	215
Bz-D-Arg-pNA	0	0	0
Ac-L-Lys-pNA	1	0.018	18
Ac-L-Tyr-pNA	1.7	0.025	15
Z-Phe-ONp	0.02	15	750,000

[a]The derivatives of Gly, Ala, and Asp were not hydrolyzed.
[b]Abbreviations: pNA = *p*-nitroanilide; ONp = *p*-nitrophenyl
ester.

phenyl alanine amide, and arginine amide. Furthermore, a thermoactive serine proteinase with maximal activity at 80°C was investigated from a newly isolated bacterium from the Azores islands. This organism grows optimally at 70°C and was
identified as *Fervidobacterium pennavorans*. The enzyme system is capable of degrading insoluble proteins which are derived from chicken feather, hair, or wool.
The keratin degrading enzyme(s) are active in a broad temperature and pH range;
50–100°C and pH 6.0–11.0.

Unlike the proteinases described above, the enzyme system of *Sulfolobus acidocaldarius*, which grows optimally at 70°C and pH 2.0, is not influenced by serine
protease inhibitors. The proteinase of *S. acidocaldarius* is active under extremely
low pH values and high temperatures, namely pH 2.0 and 90°C [14, 15].

CONCLUSION

The steady increase in the number of newly isolated thermophilic microorganisms documents the increased interest of the scientific community in hyperthermophiles. Although major advances have been made in the last decade, our knowledge
on the physiology, metabolism, enzymology, and genetics of this fascinating group
of organisms is still limited. In-depth information on the molecular properties of
the enzymes and their genes, however, has to be obtained in order to analyze the
structure and function of proteins that are functional even above 100°C. Future
research has to reveal which strategies the evolutionary distinctive archaea have
developed to ensure the remarkable thermostability of their enzymes. There is little
doubt that this group of organisms will supply novel catalysts with unique properties
that will also give a strong impetus to the development of new applications. Due to

the unusual properties of these enzymes, they are expected to fill the gap between biological and chemical industrial processes.

REFERENCES

[1]   G. Fiala and K. O. Stetter, *Arch. Microbiol., 145*, 56–61 (1986).
[2]   R. Huber, M. Kurr, H. W. Janasch, and K. O. Stetter, *Nature, 342*, 833–834 (1989).
[3]   R. Koch, A. Spreinat, K. Lemke, and G. Antranikian, *Arch. Microbiol., 155*, 572–578 (1991).
[4]   G. Antranikian, in *Microbial Degradation of Natural Products* (G. Winkelmann, Ed.), VCH, Weinheim, Germany, 1992, pp. 27–56.
[5]   R. Koch, P. Zablowski, A. Spreinat, and G. Antranikian, *FEMS Microbiol. Lett., 71*, 21–26 (1990).
[6]   S. H. Brown, H. R. Costantino, and R. M. Kelly, *Appl. Environ. Microbiol., 56*, 1985–1991 (1990).
[7]   K. Morihara and K. Oda, in *Microbial Degradation of Natural Products* (G. Winkelmann, Ed.), VCH, Weinheim, Germany, 1992, pp. 293–364.
[8]   R. Eggen, A. Geerling, J. Watts, and W. M. de Vos, *FEMS Microbiol. Lett., 71*, 17–20 (1990).
[9]   I. Blumentals, A. Robinson, and R. M. Kelly, *Appl. Environ. Microbiol., 56*, 1992–1998 (1990).
[10] H. Connaris, D. Cowan, and R. Sharp, *J. Gen. Microbiol., 137*, 1193–1199 (1991).
[11] H. Simpson, U. Haufler, and R. M. Daniel, *Biochem. J., 277*, 413–417 (1991).
[12] M. Klingeberg, F. Hashwa, and G. Antranikian, *Appl. Microbiol. Biotechnol., 34*, 715–719 (1991).
[13] D. A. Cowan, K. Smolenski, R. M. Daniel, and H. W. Morgan, *Biochem. J., 247*, 121–133 (1987).
[14] X. Lin and J. Tang, *J. Biol. Chem., 265*, 1490–1495 (1990).
[15] M. Fusek, X. Lin, and J. Tang, *Ibid., 265*, 1496–1501 (1990).

BLENDS OF BACTERIAL POLY(3-HYDROXYBUTYRATE) WITH CELLULOSE ACETATE BUTYRATE IN ACTIVATED SLUDGE

GIANPAOLO TOMASI and MARIASTELLA SCANDOLA

Dipartimento di Chimica 'G. Ciamician' dell'Università di Bologna
    and Centro di Studio per la Fisica delle Macromolecole del C.N.R.
Via Selmi 2, 40126 Bologna, Italy

ABSTRACT

Blends of bacterial poly(3-hydroxybutyrate) (PHB) with cellulose acetate butyrate (CAB) were prepared by compression molding followed by different thermal treatments. The morphology of the blends, which are totally miscible in the melt, was investigated by differential scanning calorimetry, x-ray diffraction, and optical and scanning electron microscopy. Depending on composition and thermal treatments, the blends are either single-phase amorphous PHB/CAB mixtures or partially crystalline materials composed of space-filling spherulites, where a constant fraction of the PHB present in the blend (about 65%) constitutes the lamellar crystalline phase, while the interlamellar amorphous phase is formed by the remaining PHB mixed with CAB. PHB/CAB blends over the whole composition range show no weight loss after 12 months of exposure to activated sludge. In the same experimental conditions, pure PHB quickly biodegrades at a rate which is found to be affected—at a constant degree of crystallinity—by small morphological variations. The results suggest that crystalline PHB lamellae can be attacked only after some interlamellar amorphous material has been removed. In the blends, nondegradability of interlamellar PHB/CAB mixtures prevents enzyme action toward the pure PHB lamellar crystals.

79

INTRODUCTION

Bacterial poly(3-hydroxybutyrate), PHB, and related microbial poly(hydroxyalcanoates), PHA, have attracted much attention as biodegradable and biocompatible thermoplastic polymers [1, 2]. However, wide-scale applications of bacterial polymers are still prevented by high production cost and some undesirable properties (brittleness, for example). A widespread practice in polymer science to modify physical properties is polymer blending. In the case of PHAs, blending also affects biodegradability, and a number of biodegradation studies on PHA-based blends have been recently published [3–8]. It is generally found that when PHAs form miscible blends with nonbiodegradable polymers, biodegradability substantially decreases [4–7]. However, when the two blend components are immiscible, phase separation occurs and in some cases an increase of PHA biodegradation rate has been found [3–5, 8]. In the presence of water-soluble polymers as second blend components, i.e., poly(ethylene oxide) [3] and starch [8], accelerated degradation was attributed to an increase of surface area exposed to enzymatic attack.

Recently, partial biodegradation of miscible blends of poly(3-hydroxybutyrate-co-3-hydroxyvalerate) with cellulose acetate butyrate (PHBV)/CAB in the range of PHBV-rich blends have been reported [7]. The blends were in the form of films obtained by solvent casting, a technique which was shown to induce partial segregation of the components and allows one surface to enrich in the biodegradable PHBV component. Biodegradation was found to occur preferentially on that surface. The present work investigates the biodegradability of a similar system, PHB/CAB, but the blend preparation technique is different. Compression molding instead of solution casting is used on account of good reproducibility, absence of any sidedness, and better representation of manufactured materials. Phase behavior identical to that of melt mixed and injection molded blends is obtained. It is important to point out that most results presently available in the literature on biodegradability of PHB-based blends refer to solvent-cast films.

It is known that the rate of enzymatic degradation of PHB decreases with increasing crystallinity [9, 10]. In this paper the effect of small morphological variations on PHB biodegradation is also investigated in order to show that rather subtle changes—not affecting the overall degree of crystallinity—significantly alter the rate of biodegradation of PHB.

EXPERIMENTAL

Materials

Bacterial poly(3-hydroxybutyrate), PHB, was an ICI product (G08, $M_w = 5.39 \times 10^5$, $M_w/M_n = 4.11$). Cellulose acetate butyrate was an Eastman Kodak product (EAB 500-1) whose degree of substitution ($DS_{Bu} = 2.58$, $DS_{Ac} = 0.36$) and molecular weight ($M_w = 1.30 \times 10^5$, $M_w/M_n = 2.12$) were kindly determined by Dr. C. Buchanan, Eastman Chemical Co., Kingsport, Tennessee.

Blend Preparation

In order to obtain PHB/CAB blends, weighed amounts of PHB and CAB (weight ratio 80/20, 60/40, 40/60, and 20/80) were dissolved in chloroform (5% w/v solutions). Blend films were obtained by solvent evaporation in glass Petri dishes followed by vacuum drying overnight. The solution cast blends, inserted between two aluminum plates containing a Teflon spacer, were compression molded using a Carver C12 laboratory press by heating at 190°C for 1 minute under a pressure of 2 tons/m^2. The molten blends were then subjected to one of the following thermal histories: 1) cooling to room temperature (RT), where crystallization was allowed to occur over a period of at least 2 weeks (RT blends); 2) same as in 1) followed by annealing in an oven at 90°C for 63 hours (RT + 90 blends); 3) rapid transfer into an oven at 90°C where PHB was allowed to melt-crystallize isothermally for 63 hours (IC90 blends).

Pure CAB films were obtained as described for RT blends (solvent casting, compression molding, and RT quenching). As regards PHB, the original powder was directly compression molded at 195°C for 1.5 minutes under a pressure of 2 ton/m^2, and the molten sample so obtained was subjected to the same thermal treatments (RT, RT + 90, IC90) described for the blends. The thickness of the compression-molded films was 0.11 ± 0.01 mm for blends and CAB and 0.26 ± 0.02 mm for pure PHB.

Biodegradation Experiments

The compression-molded films were cut into squares 2 × 2 cm^2. Four replicate samples of each type were carefully weighed and fastened to a fishing net (mesh size 2 × 2 mm^2) by means of a thin nylon fishing line (diameter 0.16 mm). The net was folded and sewn to form a bag which was tied to an aluminum frame. An aluminum scaffolding, holding several frames, was used to suspend the sample-containing bags in activated sludge at the municipal wastewater treatment plant of Bologna (Italy). At selected times the bags were recovered, the samples were removed, carefully washed in running water using a soft paint brush, and finally dried to constant weight under vacuum. Sterile controls (samples sterilized overnight in an atmosphere saturated with formaldehyde) were run in analogous conditions using sterilized sludge in sealed bottles. The biodegradation results are reported as weight loss divided by sample surface area ($\Delta m/S$). On account of the thinness of the films, the lateral area was neglected and S was taken as 8 cm^2. No estimate was attempted of possible surface area changes during biodegradation.

Experimental Methods

Differential scanning calorimetry was performed with a DuPont 9900 Thermal Analyzer in the temperature range −80/220°C at a heating rate of 20°/min. The temperature scale was calibrated with high purity standards. Wide-angle x-ray diffraction measurements (WAXS) were carried out with a Philips PW1050/81 diffractometer controlled by a PW1710 unit, using nickel-filtered CuKα radiation (λ = 0.1542 nm, 40 kV, 30 mA). The average crystal size was determined by the Sherrer equation from the 020 reflection, using Warren's correction for instrumental line broadening [11]. A polarizing optical microscope (Zeiss Axioscop) was used to observe sample morphology. Film surfaces were examined before and after biodegradation experiments using a scanning electron microscope (Phillips 515); samples were sputter coated with gold.

RESULTS AND DISCUSSION

Sample Characterization

Owing to the absolutely regular structure resulting from biosynthesis, PHB is a highly crystallizable polymer which develops a crystalline fraction of 60–70% in normal conditions. As commonly found in melt-crystallized polymers, PHB adopts a spherulitic morphology upon solidification from the melt. Optical microscope observations of the three PHB samples used in this work (RT, RT + 90, IC90) show spherulites of a more homogeneous and larger size in IC90 PHB (isothermally crystallized at 90°C) than in the other two samples (crystallized at room temperature). In line with previous evidence [12], this result indicates that with decreasing crystallization temperature the nucleation density increases and the spherulite size is reduced. The crystallinity degree (X_c) of the three PHB samples was obtained from DSC by comparison of the enthalpy (ΔH_m) associated with the melting endotherm with the literature value for 100% crystalline PHB [12]. The X_c values, reported in Table 1, show no variations (within the accuracy limits of this type of estimate) as a consequence of the thermal history applied to the samples. The only appreciable difference regards quality better than quantity of the crystalline phase: WAXS measurements show that the average crystal size is enhanced by annealing and by an increase of crystallization temperature. The crystal size data, reported in Table 1, are seen to increase in the order (RT) < (RT + 90) < (IC90). It is quite clear that annealing or crystallizing closer to the melting temperature of PHB ($T_m = 175°C$) would have sensibly increased the crystallinity degree, but this was out of the scope of the thermal treatments applied to the samples.

In previous papers [13–16] it was shown that bacterial PHB and a number of cellulose esters—including the CAB used in the present work—are miscible in the melt over the whole composition range. In the course of this investigation it was observed that all the compression-molded (RT) PHB/CAB films were transparent immediately after quenching from the melt to room temperature. Only the 80/20 and 60/40 blends turned opaque upon room storage due to PHB crystallization. The reason for this behavior has to be sought in the glass transition temperature (T_g) of the blends [13]. When the amount of PHB is lower than 60%, T_g is higher than room temperature and the blends are stable glassy mixtures; on the contrary, when PHB content is $\geq$ 60%, T_g lies below RT and the PHB chains have enough mobility to rearrange and crystallize. PHB-rich blends are therefore partially crys-

TABLE 1.   Degree of Crystallinity (X_c) and Average Crystal Size (C.S.) of PHB with Different Thermal Histories

Sample	X_c[a]	C.S., Å [b]
RT	0.65	208
RT + 90	0.67	252
IC90	0.66	290

[a]From DSC by comparison with the enthalpy of fusion of 100% crystalline PHB from Reference 12. $\sigma = \pm 2\%$.

[b]From WAXS using the (020) reflection. $\sigma = \pm 5\%$.

talline, being composed of a crystalline pure PHB phase and a mixed amorphous PHB/CAB phase. DSC measurements were carried out on all PHB/CAB blends. As regards the RT blends, a melting endotherm in the DSC curve confirms the visual observations, i.e., the presence of a PHB crystalline phase in both 60/40 and 80/20 blends. The enthalpy associated with the melting process (ΔH_m) is reported in Fig. 1(a) together with the value for pure PHB with the same thermal history (RT). The phase behavior of RT blends is identical to that of blends obtained by melt mixing and injection molding [13]. When RT blends are subjected to annealing at 90°C, i.e., RT + 90 blends, the 40/60 blend also turns opaque, indicating that some PHB has crystallized during permanence at 90°C. Crystallization of PHB occurs because the blend T_g, higher than room temperature, is lower than the annealing temperature. The ΔH_m values of RT + 90 and IC90 blends are reported in Figs. 1(b) and 1(c), respectively, as a function of composition. The three plots of Fig. 1 show a linear dependence of ΔH_m on composition, irrespective of the thermal treatment applied. The results indicate that a constant fraction of the PHB present in each blend segregates as a pure crystalline phase. This fraction corresponds to the X_c of pure PHB, i.e., 65–67% (see Table 1). The remaining bacterial polymer enters the mixed amorphous phase with the cellulose ester, a phase obviously richer in CAB than the overall blend composition.

X-ray diffraction measurements confirmed the calorimetric results: the crystalline phase which is seen to melt by DSC is clearly identified as PHB from the typical x-ray pattern [17, 18]. The morphology of PHB/CAB films was examined by means of polarizing optical microscopy, a technique also used to study the isothermal crystallization from the melt of the blends [19]. As expected in the absence of a crystalline phase, no birefringence was shown by any of the 20/80 blends nor by the RT (40/60) blend. In the remaining RT and RT + 90 blends, birefringent space-filling spherulites of small dimension could be better observed with higher PHB content. The two IC90 blends showed well-developed impinging spherulites.

Biodegradation

Biodegradation experiments in activated sludge of PHB/CAB blends and of the pure components were carried out over a period of 12 months. As expected, the

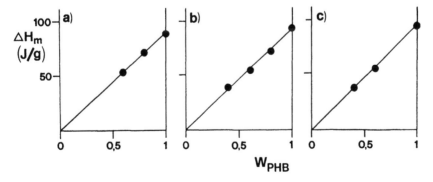

FIG. 1.   Melting enthalpy as a function of PHB weight fraction of PHB/CAB blends subjected to different thermal histories (see text): a) RT; b) RT + 90; c) IC90.

pure bacterial polyester degraded quickly, no film samples being recovered after 20–25 days of exposure. Figure 2 shows the biodegradation results of the different types of PHB (RT, RT + 90, and IC90), reported as weight loss divided by surface area ($\Delta m/S$). Upon sample recovery after any given exposure time, weight loss was found to increase invariably in the order (IC90) < (RT + 90) < (RT). Since sterile controls, examined after a period of 7 months in sterilized sludge, do not show appreciable weight loss, the observed degradation is not simple chemical hydrolysis but is an enzymatically mediated process. The results of Fig. 2 show that in PHB film samples having practically the same degree of crystallinity, small morphological variations, consisting in an increase of the average crystal size and—in the case of sample IC90—in larger and more regular spherulites, are effective in altering the rate of biodegradation. Linear regression of the weight loss data yield biodegradation rate values of 0.88, 0.72, and 0.64 mg/($\text{cm}^2\cdot$day) for (RT), (RT + 90), and (IC90) PHB, respectively.

It is well known that biodegradation of PHB is a surface phenomenon which proceeds via hydrolysis and dissolution [20]; moreover, when the polymer adopts a spherulitic morphology, degradation starts from the polymer chains in the disordered amorphous phase and proceeds to the exposed lamellar crystals [9]. In the course of the present investigation, SEM observations on degraded films of pure PHB after different times of exposure to activated sludge have confirmed that degradation proceeds sequentially and that upon erosion of the exposed amorphous phase, naked crystalline lamellae irradiating from the spherulite center clearly appear on the surface. Figure 3 shows the effect of 4 days of exposure to activated sludge of a compression-molded (RT) PHB sample. An estimate of the degree of crystallinity (by DSC) at various stages of the degradation process (up to a weight loss of 70%) shows no appreciable X_c changes, suggesting that the surface layer where the amorphous phase is preferentially degraded is thin. From SEM observa-

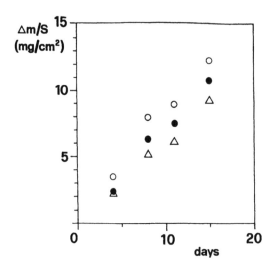

FIG. 2.   Weight loss in activated sludge as a function of time of PHB films: ($\bigcirc$) RT; ($\bullet$) RT + 90; ($\triangle$) IC90.

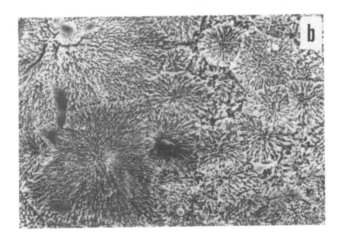

FIG. 3.   Scanning electron micrographs of the surface of a PHB (RT) film: (a) before and (b) after 4 days in activated sludge; bar = 0.1 mm.

tions of freeze-fractured biodegraded samples, the layer thickness is estimated as 5–10 μm.

The cellulose ester used in this work was an almost totally substituted cellulose (DS_{Bu} = 2.58; DS_{Ac} = 0.36), which showed no weight loss during twelve months of immersion in activated sludge. It was recently demonstrated that while films of cellulose acetate with DS $\leq$ 2.5 substantially degrade in both wastewater activated sludge [21] and aerobic composting bioreactors [21, 22], in the same experimental conditions cellulose triacetate shows no biodegradation in terms of weight loss [21]. Along these lines, a decrease of biodegradability with increasing DS in acetic esters of starch, cellulose, and xylan was recently reported by Glasser et al. [23]. It was also suggested [21] that the first step in cellulose acetate degradation is hydrolysis of the acetyl groups; only afterward do the cellulases become effective in degrading the polysaccharide main chain. As regards substituents bulkier than acetate, recently Komarek et al. [24] showed that cellulose propionate with DS = 1.8 undergoes biodegradation in an in-vitro aerobic culture system. It is reasonable to suppose that the problems connected with increasing degree of substitution in cellulose acetate will be enhanced the longer the substituent. In keeping with such considerations, the highly substituted CAB used in this work showed no weight loss during 1 year of immersion in activated sludge.

After the same period of time, no weight loss was observed in PHB/CAB blends, irrespective of the type of thermal treatment applied prior to exposure (RT, RT + 90, or IC90). For the sake of simplicity, let us consider RT blends. In the range of CAB contents of 50–100%, they are composed of a single homogeneously mixed amorphous phase, which is glassy at room temperature. The absence of weight loss in these blends indicates that the attack to PHB macromolecules by PHB depolymerases, whose existence in the environmental conditions employed is demonstrated by fast biodegradation of pure PHB, is totally inhibited by the presence of intimately mixed cellulose ester chains. When the amount of CAB component decreases below 50%, part of the bacterial polymer present in the blends segregates through crystallization and constitutes a crystalline pure PHB phase, coexisting with an amorphous mixture of varying composition. Quite surprisingly, PHB-rich blends showed no biodegradation at all, the surface of the exposed films remaining smooth (by SEM). Figure 4 illustrates the results obtained with the 80/20 RT blend together with weight loss data relative to pure (RT) PHB for the sake of comparison. A result relative to a 80/20 blend prepared by solvent casting is also included in Fig. 4 to show the effect of sidedness: a layer of practically pure PHB forms upon evaporation at the solution–air interface, which biodegrades, causing weight loss. The melting enthalpy data of Fig. 1 can be used to estimate the amount of bacterial polymer present as a segregated crystalline phase in the 80/20 RT blend. The calculations show that as much as 52% of the total blend weight is represented by crystalline PHB. It was already mentioned that in PHB-rich blends prepared by compression molding the bacterial polymer crystallizes according to a spherulitic

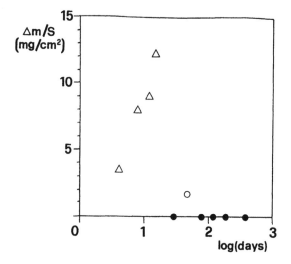

FIG. 4.   Weight loss in activated sludge as a function of time of ($\triangle$) RT PHB and of 80/20 PHB/CAB blend; ($\bigcirc$) compression molded (RT type); ($\bullet$) solvent cast.

morphology: the lamellae are formed of pure PHB and the amorphous PHB/CAB mixture is located in the interlamellar space. If 52% of the blend weight in the 80/20 sample is crystalline PHB, it is reasonable to assume that the sample surface will expose some PHB crystals to enzymatic attack. Experimental evidence of the total absence of enzymatic activity toward accessible phase-segregated PHB is puzzling and deserves some comment.

Work by Saito and coworkers [25-27] dealing with a poly(3-hydroxybutyrate)-depolymerase from *Alcaligenes faecalis* T1 proposes different mechanisms of action of the enzyme toward different substrates. In particular, it is suggested that while water-soluble 3-hydroxybutyrate oligomers are enzymatically hydrolyzed in an *endo* manner [25], a solid sample of the hydrophobic polymer is degraded in an *exo* fashion through attack to the free hydroxyl chain ends and the release of dimers [26, 27]. In the case of a PHB sample with spherulitic morphology, the *exo* degradation mechanism requires that enzymatic attack starts from the hydroxyl chain ends which are located (together with loops and trapped disordered chain segments) in the interlamellar amorphous phase, in agreement with extensive experimental evidence of initial interlamellar biodegradation. Although extension of the mechanism proposed for the *Alcaligenes faecalis* T1 depolymerase to other PHB-degrading enzymes is highly arbitrary, it is interesting to note that the present results seem to fit an analogous picture. Let us consider blend 80/20, where no biodegradation occurs during 1 year of exposure to activated sludge, notwithstanding a large—and supposedly partly accessible—pure crystalline PHB fraction. In this partially crystalline blend the hydroxyl PHB chain ends—hypothetically required by depolymerases to initiate hydrolysis—reside in a mixed interlamellar phase. With the reminder that no evidence of biodegradation is obtained in amorphous PHB/CAB blends, it appears that in blend 80/20 the lack of enzymatic attack to exposed PHB crystals is dictated by chain terminal inaccessibility. It is not yet clear why PHB in the amorphous mixed phase does not undergo enzymatic hydrolysis.

One hypothesis contemplates a dramatic decrease of chain mobility with respect to pure PHB. An alternative reason might be a detrimental change of surface hydrophobicity brought about by the partner polymer in the blend. In this respect it was demonstrated [28, 29] that enzymatic degradation of solid PHB requires depolymerases containing a hydrophobic domain which provides the binding site to the substrate. The cellulose ester intimately mixed with PHB might modify the hydrophobic character of the surface, thus preventing enzyme attachment. Both possibilities (loss of amorphous phase mobility and change of hydrophobicity of the polymeric substrate) need to be tested by selecting appropriate polymers to be blended with PHB. Work is in progress along these lines.

CONCLUSIONS

The present work has shown that rather subtle morphological changes in spherulitic bacterial PHB—varying the average crystal size at constant overall crystallinity—affect the rate of biodegradation in activated sludge. This result calls for attention when comparing data from different sources in the absence of a detailed morphological characterization. The lack of any experimental evidence of biodegradation in all PHB/CAB blends exposed for 12 months to activated sludge raises a number of questions. Most striking in the absence of enzymatic attack to blend 80/20, where pure crystalline PHB amounts to more than 50% of the total blend weight. In pure spherulitic PHB, the depolymerases attack the accessible interlamellar amorphous phase at first, but eventually the crystalline lamellae also undergo biodegradation. The surface layer being sequentially degraded is thin and does not change much in the course of biodegradation, indicating that once the lamellae are cleared of the interlamellar material, they are easily attacked by the enzymes. Apart from possible effects on enzyme—substrate interaction due to changes in surface hydrophobicity upon blending, the results of the present work seem to suggest that attack to the lamellar PHB crystalline phase in crystalline PHB/CAB blends is conditioned by prior consumption of some interlamellar amorphous material. Nondegradability of the mixed interlamellar phase seems to be responsible for inaccessibility to the enzyme of the pure bacterial polymer phase represented by the crystalline lamellae.

REFERENCES

[1] P. A. Holmes, in *Developments in Crystalline Polymers* (D. C. Basset, Ed.), Elsevier, New York, 1988, Vol. 2.
[2] Y. Doi, *Microbial Polyesters*, VCH Publishers, New York, 1990.
[3] Y. Kumagai and Y. Doi, *Polym. Degrad. Stab., 35*, 87 (1992).
[4] Y. Kumagai and Y. Doi, *Ibid., 36*, 241 (1992).
[5] Y. Kumagai and Y. Doi, *Ibid., 37*, 253 (1992).
[6] P. Sadocco, C. Bulli, E. Elegir, A. Seves, and E. Martuscelli, *Makromol. Chem., 194*, 2675 (1993).
[7] D. F. Gilmore, R. C. Fuller, B. Schneider, R. W. Lenz, N. Lotti, and M. Scandola, *J. Environ. Polym. Degrad.*, In Press.
[8] B. A. Ramsay, V. Langlade, P. J. Carreau, and J. A. Ramsay, *Appl. Environ. Microbiol., 59*, 1242 (1993).
[9] Y. Kumagai, Y. Kanesawa, and Y. Doi, *Makromol. Chem., 193*, 53 (1992).
[10] M. Parikh, R. A. Gross, and S. P. McCarthy, *Polym. Mater. Sci. Eng., 66*, 408 (1992).
[11] H. P. Klug and L. E. Alexander, *X-Ray Diffraction Procedures for Polycrystalline and Amorphous Materials*, Wiley, New York, 1974.
[12] P. J. Barham, A. Keller, E. L. Otun, and P. A. Holmes, *J. Mater. Sci., 19*, 2781 (1984).
[13] M. Scandola, G. Ceccorulli, and M. Pizzoli, *Macromolecules, 25*, 6441 (1992).
[14] N. Lotti and M. Scandola, *Polym. Bull., 29*, 407 (1992).
[15] G. Ceccorulli, M. Pizzoli, and M. Scandola, *Macromolecules, 26*, 6722 (1993).
[16] M. Pizzoli, M. Scandola, G. Ceccorulli, and U. Piana, *Book of Abstracts of the 4th European Symposium on Polymer Blends*, Capri, 1993, p. 309.
[17] K. Okamura and R. H. Marchessault, in *Conformation in Biopolymers* (Ramachandran, Ed.), Academic Press, London, 1974, pp. 709-720.

[18]   S. Brückner, S. V. Meille, L. Malpezzi, A. Cesàro, L. Navarini, and R. Tombolini, *Macromolecules, 21*, 967 (1988).

[19]   M. Pizzoli, M. Scandola and G. Ceccorulli, *Ibid.*, Submitted.

[20]   Y. Doi, Y. Kanesawa, M. Kunioka, and T. Saito, *Ibid., 23*, 26 (1990).

[21]   C. M. Buchanan, R. M. Gardner, and R. J. Komarek, *J. Appl. Polym. Sci., 47*, 1709 (1993).

[22]   J.-D. Gu, D. T. Eberiel, S. P. McCarthy, and R. A. Gross, *J. Environ. Polym. Degrad., 1*, 143 (1993).

[23]   W. G. Glasser, G. Ravindran, G. Samaranayake, R. K. Jain, J. S. Todd, and J. Jervis, Paper Presented at the 206th ACS National Meeting, Division of Cellulose, Paper and Textile, Chicago, 1993.

[24]   R. J. Komarek, R. M. Gardner, C. M. Buchanan, and S. Gedon, *J. Appl. Polym. Sci., 50*, 1739 (1993).

[25]   Y. Shirakura, T. Fukui, T. Saito, Y. Okamoto, T. Narikawa, K. Koide, K. Tomita, T. Takemasa, and S. Masamune, *Biochim. Biophys. Acta, 880*, 46 (1986).

[26]   T. Tanio, T. Fukui, Y. Shirakura, T. Saito, K. Tomita, T. Kaiho, and S. Masamume, *Eur. J. Biochem., 124*, 71 (1982).

[27]   J. J. Jesudason, R. H. Marchessault, and T. Saito, *J. Environ. Polym. Degrad., 1*, 89 (1993).

[28]   T. Fukui, T. Narikawa, K. Miwa, Y. Shirakura, T. Saito, and K. Tomita, *Biochim. Biophys. Acta, 952*, 164 (1988).

[29]   K. Mukai, K. Yamada and Y. Doi, *Int. J. Biol. Macromol., 15*, 361 (1993).

BIODEGRADATION OF CELLULOSE ESTERS: COMPOSTING OF CELLULOSE ESTER–DILUENT MIXTURES

CHARLES M. BUCHANAN,* DEBRA D. DORSCHEL,
ROBERT M. GARDNER, RON J. KOMAREK, and ALAN W. WHITE

Research Laboratories
Eastman Chemical Company
P.O. Box 1972, Kingsport, Tennessee 37662

ABSTRACT

A number of polymers such as polylactic acid (PLA), polycaprolactone (PCL), polyhydroxybutyrate (PHB), Matter-Bi, cellulose acetate (CA) with different degrees of substitution (DS), and cellulose ester–diluent mixtures have been evaluated in a static, bench-scale simulated municipal compost environment. Of the polymers evaluated, cellulose acetate (DS < 2.2), poly(hydroxybutyrate-*co*-valerate) (PHBV), and PCL exhibited the fastest composting rates, completely disappearing after 14 days. Optically clear resins were prepared from CA (DS = 2.06) and triethylcitrate (TEC) by thermal compounding, and the resins were converted to compression-molded film and injection-molded bars for composting studies. A series of miscible blends consisting of cellulose acetate propionate (CAP) and poly(ethylene glutarate) (PEG) or poly(-tetramethylene glutarate) (PTG) were also prepared and evaluated in composting. In addition to measured weight loss, samples were removed from the compost at different intervals and evaluated by gel permeation chromatography and ^{1}H NMR. As expected, the CA/TEC films disappeared rapidly upon composting while the injection-molded bars exhibited weight losses of 10–12%. For the CAP/polyester blends, the type of polyester (PEG versus PTG) in the blend made no difference in composting rates. In general, as the DS of the CAP decreased and the amount of polyester in the blend increased, the rate of composting and the weight loss due to composting increased. When the CAP was highly substituted, almost all of the weight loss was ascribed to loss to polyester. When the DS of the CAP was below approximately 2.0, both components degraded.

INTRODUCTION

The concept of biodegradable polymers as a solution to the disposal of solid waste is certainly not new. In a series of patents and reports in the early 1970s, Potts et al. demonstrated that polycaprolactone (PCL) was a suitable biodegradable polymer for many applications [1]. Potts et al. also showed that PCL was miscible or partially miscible with a wide range of the other polymers, which opened the door to the use of polymer blends or composites as biodegradable materials. In part, the interest in biodegradable polymers during this period of time was driven by the oil embargo and the expected shortage of petroleum as a feedstock for synthetic polymers. Over time, the fear of oil shortages waned. Furthermore, due to material cost and deficiencies, biodegradable polymers never fulfilled the expectations and hopes of that period. Consequentially, interest in biodegradable polymers faded. In the late 1980s it began to become apparent to many that available landfill space was becoming scarce in many metropolitan areas of the United States and Europe, leading to a renewed interest in biodegradable polymers and the use of composting, along with recycling and landfill, as a component for the solution of the emerging solid waste disposal crisis. Much of the research in the period from 1989 to the present has focused on starch or starch composites as new biodegradable alternatives to traditional synthetic polymers [2]. Other polymers, including polyalkanoates [e.g., polyhydroxybutyrate (PHB)] as a typical example, have also received considerable attention [3]. Very recently, it has been proposed that cellulose esters, e.g., cellulose acetate (CA), be considered as potentially useful polymers in biodegradable applications [4–11].

Historically, there has been considerable confusion regarding the biodegradation potential of cellulose esters. It was generally accepted that cellulose esters with a degree of substitution less than 1.0 will degrade from attack of microorganisms at the unsubstituted residues of the polymers, and that the ether linkages to the cellulose backbone are generally resistant to microbial attack [12]. Work by Stutzenberger and Kahler indicated that CA was a poor substrate for microbial attack [13]. Conflicting evidence was provided by Reese who isolated cellulolytic filtrates which deacetylated soluble CA (DS = 0.76) and insoluble cellobiose octaacetate [14]. Additional early evidence as to the biodegradation potential of CA was provided by Cantor and Mechalas [15]. Cantor and Mechalas demonstrated that reverse-osmosis membranes prepared from CA with a DS = 2.5 suffered loses in semipermeability due to microbial attack. Since 1992 there have been several reports which clearly demonstrate that CA having a DS of less than approximately 2.5 is inherently biodegradable under both aerobic [4, 6] and anaerobic [7] conditions, and that many cellulose acetates are readily compostable [5, 8–11]. A general finding was that as the DS of the CA decreased, the rate of biodegradation increased; below a DS of ca. 2.1, composting rates of CA approached or exceeded that of many other commercial biodegradable polymers. Very recently, Komarek et al. showed, via aerobic biodegradation of radiolabeled cellulose propionates, that cellulose propionates below a DS of ca. 1.85 are also potentially useful as biodegradable polymers [6]. This work indicates that as the DS and the length of the acyl side group decreases, the rate of biodegradation increases; longer adaptation periods, relative to CA, were also required for the cellulose propionates.

Although careful consideration has been given to the biodegradation of cellulose esters, scarce attention has been paid to the biodegradation of formulated resins consisting of cellulose esters and diluents. This is no small matter as the melt processing temperature of most cellulose esters exceeds that of the decomposition temperature of the cellulose esters, which implies that most cellulose esters must be plasticized if they are to be used in thermoplastic applications [16]. Examining the biodegradation of a formulation consisting of two or more components imparts a new level of complexity to the problem. With polymer–diluent mixtures, the question of morphology becomes paramount. Also, with homogeneously mixed polymer–diluent or polymer–polymer mixtures, the interesting question of how the biodegradation rate of one component influences the biodegradation rate of the other component is raised.

In this account we describe the application of bench-scale composting methodology developed in our laboratories to a number of commercially available polymers that are currently being evaluated in many laboratories. This work provides a benchmark of the relative composting rates expected for these polymers in an efficient municipal window composting operation. We also report some preliminary work involving the application of this composting methodology to cellulose acetate–triethylcitrate (polymer–diluent) mixtures and to cellulose acetate propionate–polyester (polymer–polymer) blends.

MATERIALS AND METHODS

L-Polylactic acid (PLLA) and 1/1 D,L-polylactic acid (PDLLA) was obtained from Birmingham Polymers. Poly(hydroxybutyrate) (PHB) and poly(hydroxybutyrate-co-valerate [90/10]) (PHBV) are commercially available from Zeneca Bio-Products. Polycaprolactone (PCL) was purchased from Union Carbide, and Matter-Bi (Matter-Bi is presumably a blend of starch or modified starch and other polymers such as polyvinyl alcohol and polyacrylics [2b]) is commercially available from Novamont. Cellulose triacetate (CTA, DS = 2.97) and cellulose acetate CA398-30 (DS = 2.52) are commercially available and were obtained from Eastman Chemical Company. The remaining cellulose acetates were obtained by aqueous hydrolysis of either CTA or the CA having a DS of 2.52 [17]. Cellulose acetate propionate, having a DS of 2.75 ($DS_{Ac} = 0.01$, $DS_{Pr} = 2.74$), is commercially available from Eastman Chemical Company (Kingsport, Tennessee). The remaining cellulose acetate propiomates were obtained by acid-catalyzed, aqueous hydrolysis of this DS 2.75 CAP [17]. All of these CAPs had a very low acetyl content ($DS_{Ac} = 0.01$–0.05) and can be considered to be essentially cellulose propionates. Poly(ethylene glutarate) (PEG) and poly(tetramethylene glutarate) (PTG) were prepared by the condensation polymerization of dimethyl glutarate and the appropriate diol using $Ti(O^iPr)_4$ as the catalyst.

First scan DSC heating curves were obtained for the cellulose acetates using a DuPont 912 differential scanning calorimeter spectrometer. The samples were heated from 25 to 300°C at a heating rate of $20° \cdot min^{-1}$. The reported melting temperatures are nonequilibrium melting temperatures.

Films were prepared by solvent casting and by thermal processing. With the exception of CTA, the solvent for the cellulose acetates (20% solids) was acetone containing 3–7% water. The solvent for CTA, PHBV, and PCL was $CHCl_3$. The solution was poured onto a metal plate and a draw blade with a 15-mil well was used to give a thin film. PHB, PLLA, and PDLLA films were prepared by compression molding. Matter-Bi film was prepared by a blown film process. In all cases involving blends, the mixtures were prepared by thermal compounding and characterized according to previously described methods [16, 18, 19]. The cellulose acetate-triethylcitrate (CA/TEC) compositions were similarly prepared. The film thickness ranged from 2 to 8 mil. Injection-molded parts were 122 mil thick and were obtained as previously described [20].

Two types of composting studies were generally conducted. In one case, each compost unit contained a minimum of 15 test films and 10 Matter-Bi films which served as an internal positive control. Results reported herein are average values for all films in each individual unit. With the exception of PCL and, in one case, PLA, the films were added to the composting unit along with the synthetic compost mixture. Because of the low melting temperature of PCL, the PCL films were not added until completion of the thermophilic phase. Also, due to the relative low T_g of PLA, two studies were conducted. In one study, the PLA films were added at the beginning of the composting cycle (30-day experiment) and, in the second, the PLA films were added at the end of the thermophilic phase (24-day experiment). PHB, PHBV, and PCL were evaluated over a 14-day composting period. The cellulose acetates having a DS of 1.74, 1.86, 2.06, 2.21, 2.52, and 2.97 were evaluated over either a 14-day or a 30-day composting cycle. In the second type of study, each compost unit contained 30–35 films. Three to five films were removed every 3–5 days and characterized.

The details of the compost unit design, the control system, the compost mixture, compost characterization, and methods for characterization of samples removed from the compost have been extensively described elsewhere [5].

RESULTS AND DISCUSSION

General Composting

The primary criteria that we use in screening polymers in composting are initial breakup of the film within the first few days and total weight loss. Unless a film fractures into small enough pieces that will permit their passage through a screening operation commonly employed in many municipal composting operations, the films will be removed by the screening operation and will ultimately be landfilled. This, of course, defeats the entire purpose of developing a compostable polymer. Film breakage must be followed by complete mineralization of the fragments, and total weight loss is often a good indicator of mineralization. We generally make no effort to distinguish between chemical degradation and biological degradation in our compositng methodology. Undoubtedly, with certain substrates, both chemical and biological degradation coexist. In fact, we believe that one can take advantage of the changes in compost pH and temperature in designing polymers so that these processes will aid in the fragmentation and ultimate biological degradation of the polymeric substrate. Ultimately, complete mineralization of the substrate must be demonstrated but that is best accomplished using respirometry and radiochemical labeling techniques [4, 6]. Hence, the methodology described in this account models the first step of composting and serves as a predictor of ultimate mineralization.

Figure 1 provides the % weight loss for some of the potentially biodegradable polymers we have evaluated in our laboratories. The number of days given in parentheses is the length of the composting cycle, and the value following the CA samples is the total DS for the CA. The first two entries are for PLA films which were added to the compost after completion of the thermophilic phase (composting cycle = 24 days), while the next two entries are for the PLA films which were present in the compost for the entire composting cycle (composting cycle = 30 days). The experiment was conducted in this manner due to the relative low T_g (ca. 55°C) of the PLA which is exceeded during the thermophilic phase. As can be seen, in the case where the PLA did not experience the temperatures associated with the peak thermophilic phase (60–65°C, vide infra), the weight loss was minimal. When the PLA was present for the entire composting cycle, the % weight loss increased. Interestingly, in the latter case the PDLLA films showed over twice the % weight loss observed for the PLLA films which we believe is likely due to the fact that the PDLLA was entirely amorphous while the PLLA is highly crystalline. In both cases the PLA films remained intact, i.e., they did not break up into smaller fragments. As the next entry illustrates, PCL (added after the thermophilic phase) completely disintegrated (no pieces could be recovered following an extensive screening operation). Matter-Bi was a particularly interesting case. Approximately 50% weight loss

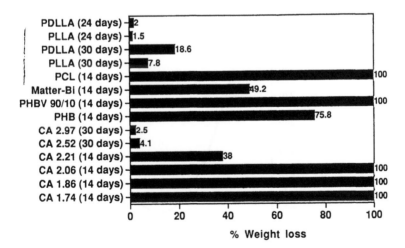

FIG. 1.   The % weight loss observed for selected potentially compostable polymers in a bench-scale simulated composting environment. The numbers in parentheses reflect the length of the composting cycle while the numbers following CA indicate the degree of substitution.

was achieved very rapidly, after which an intact, flexible film remained. When the composting cycle was extended to 180 days, no changes in the remaining Matter-Bi film was observed. Iodine staining demonstrated that the starch component was removed very rapidly, leaving a synthetic polymer residue which was very slow to degrade [5]. As expected, both the PHBV and PHB films showed extensive weight loss. The remaining examples illustrate the effect of DS on composting rates of CA. For the CAs having a DS of 2.97 and 2.52, minimal weight loss was observed and the films remained intact. The film prepared from the CA with a DS of 2.21 underwent extensive breakage and discoloring and had a % weight loss of 38% after 14 days. Upon lowering the DS to 2.06, the films completely disintegrated and no pieces could be recovered from the compost.

The time frame for complete disintegration and mineralization of a polymer during a composting operation is still subject to considerable debate. The likely outcome is that composting rates will be related to natural materials, e.g., oak leaves, under very controlled conditions. However, the data presented above illustrate that PCL, PHB, and CA (DS < 2.2) are excellent candidates as compostable polymers. The extent of their use for these applications will likely depend upon their physical properties and cost.

Cellulose Acetate/Triethylcitrate

Although cellulose acetate having a DS less than ca. 2.2 shows good composting rates, the question remains regarding the ability of the material to be thermally processed. As Kamide has shown and as our data indicate [21], when the T_m of cellulose acetate is plotted versus the DS, a minimum is observed near a DS of 2.3 (Fig. 2). Extrapolating horizontally across the minimum from a DS of 2.5 (common

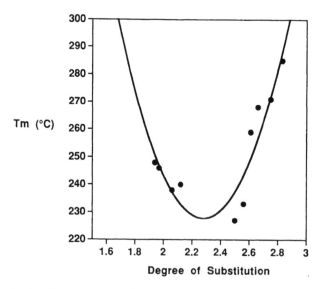

FIG. 2.   A plot illustrating the change in nonequilibrium melting temperature as a function of cellulose acetate degree of substitution.

thermoplastic CA) to a lower DS suggests that a CA with a DS of 2.0 would be thermally processible. Indeed, we have found that thermal processing of CAs in the range of 2.0 to 2.2 to be very comparable to that of CA with a DS of 2.5.

To evaluate the compostability of CA/TEC mixtures, we prepared compression-molded films (28 and 30% TEC) as well as injection-molded bars (30–31% TEC) from CA with a DS of 2.05. In the case of the bars, three cellulose acetates with identical DS but differing in molecular weight were used. These materials were then subjected to our bench-scale composting. Figure 3 shows a typical compost temperature and % H_2O profile for a 30-day composting experiment. As can be seen, the peak thermophilic phase was reached in ca. 24 hours where it remained for ca. 2 days. The temperature then dropped over a period of 5 days to near 38°C. This maturation phase was then maintained for an additional 22 days. During the thermophilic phase, the % H_2O dropped to near 58% due to the secondary air flow used to cool the units. During the maturation phase with a lesser air flow, water was produced which slowly brought the water content back near 62–63% where it remained for the life of the experiment.

Figure 4 shows the % weight loss for the films and for the bars. In the case of Entry 5, the films extensively broke up and after 12 days the experiment was terminated as only very small, discolored chips could be found. For Entry 4, the film samples also extensively fragmented but samples were removed every 4–6 days for characterization. These removed samples, which were not subjected to the full composting cycle of 30 days, were included in the final weight loss which lowers the

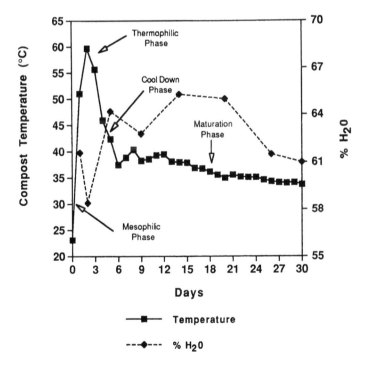

FIG. 3.   A typical compost temperature profile and compost water content for composting experiments involving cellulose acetate/triethylcitrate.

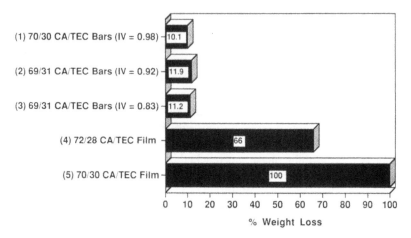

FIG. 4.   The % weight loss observed for CA/TEC film and bars. The % weight loss for Entry 4 reflects the weight of samples removed at different time intervals during the composting experiment, and thus this apparent weight loss is the lower level of % weight loss.

apparent final weight loss. In the case of the injection-molded bars, the % weight loss was very similar but, as Fig. 5 illustrates, % weight loss does not adequately describe the results. Bars, which were not subjected to composting, were washed using our normal washing procedure [5]. In this case we found that 1.2, 0.3, and 0.0% TEC was removed from the bars having an IV of 0.98, 0.92, and 0.83, respectively. This behavior is likely related to the molecular weight dependence of diluent compatibility with a polymer. That is, TEC will diffuse more rapidly from the higher molecular weight CA. If one subtracts the TEC lost from the total % weight loss and assumes that the remaining weight loss is due to biodegradation of the CA, one can attempt to calculate the weight loss normalized to the total surface area [9]. In doing this, one finds that the bar prepared from the highest molecular weight CA has nearly twice the weight los ($48.7 \ \mu g/mm^2$) of that observed for the bar prepared from the lowest molecular weight CA ($26.4 \ \mu g/mm^2$). In the case of the middle molecular weight CA, the amount of TEC lost is greater than the total weight loss, indicating that diffusion of TEC from the bars is modified after composting! Clearly, unless the diluent is completely or nearly completely removed from the molded object, normalized weight loss cannot be used for polymer–diluent mixtures. In cases involving polymer–polymer blends where diffusion is not observed, this issue presents no difficulties.

As noted, film samples were removed for further characterization. Figure 6 shows the average DS and % TEC determined by ^{1}H NMR. As can be seen, the % TEC drops rapidly to below 1.5% while the measured DS declines more slowly over 14 days to near 1.7, dropping to 1.41 at 30 days. As we and others have noted [4, 9], degradation of cellulose acetate occurs at the surface of the substrate so that bulk values such as DS reflect the composition of the unchanged interior and the constantly changing surface. A profile such as that seen in Fig. 6 is typical of what we have observed many times for CA. Finally, Fig. 7 shows the molecular weight profile for these same samples as determined by GPC. In this case, M_w is observed

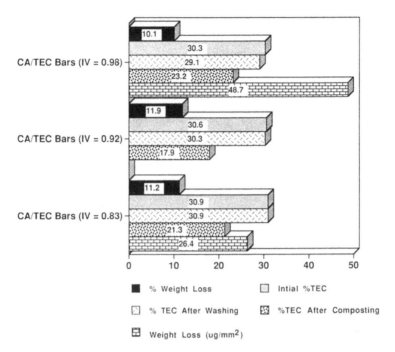

FIG. 5.   The % weight loss, initial and final TEC content, and normalized weight loss for CA/TEC bars.

to decrease while M_n correspondingly increases. This would suggest preferential internal chain cleavage as opposed to cleavage from the chain end.

Cellulose Acetate Propionate/Aliphatic Polyester Blends

In a series of recent reports we have shown that CAP forms miscible blends with a variety of aliphatic polyesters such as poly(ethylene glutarate) (PEG) or poly(tetramethylene glutarate) (PTG) [16, 18, 19]. In general, these blends are stable, optically clear, amorphous polymer blends. As we have noted above, cellulose acetate propionates with a suitable low DS are expected to be biodegradable at a reasonable rate while highly substituted CAPs are expected to biodegrade very slowly. Linear, straight-chain, aliphatic polyesters, similar to those described in this account, have been demonstrated to be biodegradable [22]. Given this background, the interesting questions which naturally arise are by what mechanism do homogeneous mixed blends of these two polymers degrade and does the biodegradation of one polymer component influence the biodegradation of the other.

In this brief account we focus on the weight loss, molecular weight, and blend composition as a function of CAP DS, polyester type (PEG vs PTG), and polyester content of the blend. The percent weight loss for these blends are shown in Fig. 8 along with entries for CAP/PHB and CAP/PDDLA blends. The observed general trend is an increasing percent weight loss as the amount of polyester in the blend increases and as the DS of the CAP decreases. Little difference in weight loss is observed between the CAP/PEG and CAP/PTG blends at the same DS and polyes-

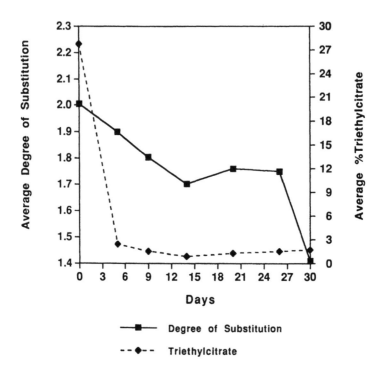

FIG. 6. The change in DS and % TEC content of CA/TEC films removed and analyzed by ^{1}H NMR during a 30-day composting experiment.

ter content. For example, Entries 7 and 8, which differ only in polyester type and which were tested in the same composting experiment, show virtually no difference in weight loss. When the CAP DS is decreased to 2.15 and the weight percent of polyester is increased by 5%, the weight loss increases by 5-10%. Maximum weight loss was observed when the DS of the CAP was lowered slightly and when the PEG became the major component in the blend. Some variation is observed between composting experiments.

Of particular interest are the differences in weight loss between the CAP/PHB blend (Entry 4), the CAP/PDLLA blend (Entry 5), and the equivalent CAP/PEG, PTG blends (Entries 6-9). Although the CAP/PHB blend has a higher content of PHB relative to the CAP/PEG, PTG blends, only 5% weight loss is observed after a 30-day composting cycle for the CAP/PHB blend versus a 20-25% weight loss observed for the CAP/PEG, PTG blends (cf. Entries 5-8) in a 10-15-day composting cycle. For the CAP/PDLLA blend, a 10% weight loss is observed which is half of the equivalent CAP/PTG blend (Entry 8) where a 20% weight loss was observed. In the past, we and others have proposed that the initial step in the biodegradation of cellulose esters involves attachment of enzyme(s) to the substrate. Biologically (esterase) mediated hydrolysis of the acetyl substitutent provides chains or chain segments with no or little substitution which can then be degraded by cellulases by a process that has been extensively studied [23]. By this mechanism, biodegradation occurs on the surface of the substrate with the effect of surface erosion. With the exception of PHB, which degrades by a mechanism similar to cellulose [24], it is

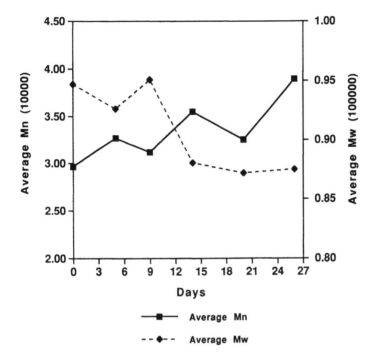

FIG. 7.   The change in number-average and weight-average molecular weight for CA/ TEC films removed and analyzed by GPC during a 30-day composting experiment.

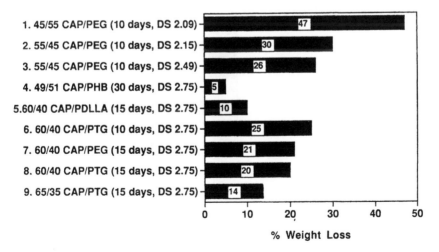

FIG. 8.   Weight loss for blends of cellulose acetate propionate and aliphatic polyesters after composting. The initial ratio provides the blend compositions. The length of the composting cycle in days and the DS of the CAP are provided within the parentheses.

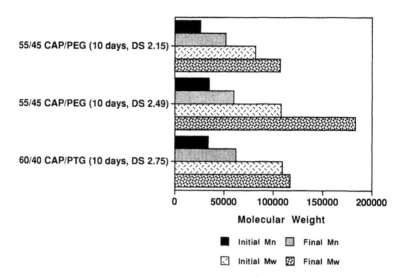

FIG. 9.   The change in weight-average and number-average molecular weights for selected CAP/aliphatic polyester blends after composting.

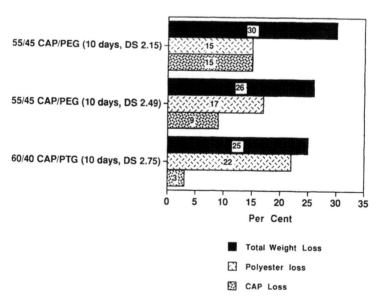

FIG. 10.   The total % weight loss, polyester loss, and CAP loss for CAP/aliphatic polyester blends after composting.

generally believed that polyesters are first hydrolytically degraded to monomers or short-chain oligomers which are then assimilated by microorganisms [22]. Most likely, for aliphatic polyesters, both chemical and biological hydrolysis occur concurrently. In the case of chemical hydrolysis, prior attachment of enzyme to substrate is not required and hydrolytic (nonbiological) degradation can occur both on the surface and in the interior of the substrate. Given this background, this would suggest that the weight loss in the PEG or PTG blends containing highly substituted CAP is probably due to initial hydrolysis of the PEG or PTG.

Figure 9 provides the initial and final molecular weights for three blends from a 10-day compost experiment. In all cases both the number-average and weight-average molecular weights increased. This would be consistent with preferential loss of the lower molecular weight polyester. A better understanding of what is occurring with these blends can be obtained from Fig. 10 where the total % weight loss is plotted versus the weight loss of polyester from the blend (determined by ^{1}H NMR). If only polyester was degrading while the CAP remained intact, then the % weight loss and the polyester loss should be equal. For the blend involving the DS = 2.75 CAP, the polyester loss and the total weight loss are nearly equal. However, at a DS of 2.15 only half of the weight loss can be ascribed to the polyester. This would indicate that for the highly substituted CAP, the polyester is degrading rapidly with little loss of CAP, while both blend components are degrading at lower degrees of substitution.

CONCLUSION

In this work we have evaluated in our bench-scale composting system many polymers which have been proposed for use as compostable polymers. Cellulose acetate (DS < 2.2), PHBV, and PCL exhibited the fastest composting rates, completely disappearing after 14 days. Each of these materials has different strengths and weakness. PLA has many desirable physical properties and has good potential if it can be made successfully on an industrial scale. We were somewhat surprised at the composting rate of our PLA relative to PHB or PCL. Matter-Bi is presently commercially available and has found use in different applications. The issue of mineralization of the nonstarch components of Matter-Bi needs to be resolved. Polyalkanoates, e.g., PHB, are generally produced commercially by fermentation. The most notable properties of PHB are its good barrier properties and excellent composting rates while the major shortcomings of PHB are its low T_g and slow crystallization rates. Both PCL and CA are produced on a larger scale than PHB and, at least for the near term, can be obtained at a lower cost. Polycaprolactone has many excellent properties. The primary problem with PCL is its low T_m which somewhat limits its use. Cellulose acetate, perhaps one of the oldest thermoplastics known, has been used in a very broad range of applications and its properties are well understood. The major difficulty with cellulose esters is that they generally must be plasticized in order to be thermally processed without decomposition or loss of physical properties.

In view of the need for placticization of cellulose esters in thermoplastic applications, we have presented preliminary studies involving CA/TEC and CAP/aliphatic polyesters. CA/TEC films disappeared rapidly upon composting while the injection-molded bars exhibited weight losses of 10–12%. For the CAP/aliphatic polyester blends, the type of polyester (PEG versus PTG) in the blend made no difference in composting rates. In general, as the DS of the CAP decreased and the amount of polyester in the blend increased, the rate of composting and the weight loss due to composting increased. When the CAP was highly substituted, almost all of the weight loss was ascribed to loss of polyester. When the DS of the CAP was below approximately 2.0, both components were observed to degrade. The observed weight loss due to composting for CAP/polyalkanoates (PDLLA and PHB) blends was less relative to the CAP/PEG, PTG blends. This work also points out some of the difficulties surrounding the study of the biodegradation of compositions containing two or more components.

REFERENCES

[1]   (a) R. A. Clendinning, J. E. Potts, and W. D. Niegisch, US Patent 3,844,987 (1974). (b) R. A. Clendinning, J. E. Potts, and W. D. Niegisch, US Patent 3,850,862 (1974). (c) R. A. Clendinning, J. E. Potts, and W. D. Niegisch, US Patent 3,850,863 (1974). (d) R. A. Clendinning, J. E. Potts, and W. D. Niegisch, US Patent 3,932,319 (1976). (e) J. E. Potts, R. A. Clendinning, W. B. Ackart, and W. D. Niegisch, *Polym. Sci. Technol., 3*, 61 (1973).

[2]   (a) G. Lay, B. Bellingen, J. Rehm, R. F. Stepo, M. Thoma, J.-P. Sachetto, D. J. Lentz, and J. Silbiger, US Patent 5,095,054 (1992). (b) C. Bastioli, V. Bellotti, G. Del Tredici, R. Lombi, A. Montino, and R. Ponti, International Publication WO 92/19680 (1992).

[3]   (a) M. Yasin and B. J. Tighe, *Plas., Rubber, Compos. Process. Appl., 19*, 15 (1993). (b) H. Nishida and Y. Tokiwa, *J. Environ. Polym. Degrad., 1*, 65 (1993). (c) Y. Kumagai and Y. Doi, *Makromol. Chem., Rapid Commun., 13*, 179 (1992). (d) Y. Kumagai and Y. Doi, *Polym. Degrad. Stab., 35*, 87 (1992). (e) Y. Kumagai and Y. Doi, *Ibid., 36*, 241 (1992). (f) Y. Kumagai and Y. Doi, *Ibid., 37*, 253 (1992). (g) J. E. Kemnitzer, S. P. McCarthy, and R. A. Gross, *Macromolecules, 25*, 5927 (1992).

[4]   C. M. Buchanan, R. M. Gardner, and R. J. Komarek, *J. Appl. Polym. Sci., 47*, 1709 (1993).

[5]   R. M. Gardner, C. M. Buchanan, R. J. Komarek, D. D. Dorschel, C. Boggs, and A. W. White, *Ibid., 52*, 1477 (1994).

[6]   R. J. Komarek, R. M. Gardner, C. M. Buchanan, and S. C. Gedon, *Ibid., 50*, 1739 (1993).

[7]   C. J. Rivard, W. S. Adney, M. E. Himmel, D. J. Mitchell, T. B. Vinzant, K. Grohmann, L. Moens, and H. Chum, *Appl. Biochem. Biotech., 34/35*, 725 (1992).

[8]   J.-D. Gu, M. Gada, G. Kharas, D. Eberiel, S. P. McCarthy, and R. A. Gross, *J. Am. Chem. Soc., Polym. Prepr.*, p. 351 (1992).

[9]   J.-D. Gu, D. Eberiel, S. P. McCarthy, and R. A. Gross, *J. Environ. Polym. Degrad., 1*, 143 (1993).

[10] R. A. Gross, J.-D. Gu, D. Eberiel, M. Nelson, and S. P. McCarthy, in *Fundamentals of Biodegradable Materials and Packaging* (D. Kaplan, E.

Thomas, and C. Ching, Eds.), Technomic Publishing Co., Lancaster, Pennsylvania, 1993.

[11]   C. M. Buchanan, R. M. Gardner, R. J. Komarek, S. C. Gedon, and A. W. White, *Ibid.*, p. 133.

[12]   (a)   U. Kasulke, H. Dautzenberg, E. Polter, and B. Philipp, *Cellul. Chem. Technol., 17*, 423 (1983). (b)   A. S. Perlin and S. S. Bhattacharjee, *J. Polym. Sci., Part C, 36*, 506 (1971). (c) M. G. Wirick, *J. Polym. Sci., Part A-1, 6*, 1965 (1968).

[13]   F. Stutzenberger and G. Kahler, *J. Appl. Bacteriol., 61*, 225 (1986).

[14]   E. T. Reese, *Ind. Eng. Chem., 49*, 89 (1975).

[15]   P. A. Cantor and B. J. Mechalas, *J. Polym. Sci., Part C, 28*, 225 (1969).

[16]   C. M. Buchanan, S. C. Gedon, A. W. White, and M. D. Wood, *Macromolecules, 26*, 2963 (1993).

[17]   C. M. Buchanan, K. J. Edgar, J. A. Hyatt, and A. K. Wilson, *Ibid., 24*, 3050 (1991).

[18]   C. M. Buchanan, S. C. Gedon, B. G. Pearcy, A. W. White, and M. D. Wood, *Ibid., 26*, 5704 (1993).

[19]   C. M. Buchanan, S. C. Gedon, A. W. White, and M. D. Wood, *Ibid., 25*, 7373 (1992).

[20]   A. W. White, C. M. Buchanan, B. G. Pearcy, and M. D. Wood, *J. Appl. Polym. Sci., 52*, 525 (1994).

[21]   K. Kamide and M. Saito, *Polym. J., 17*, 919 (1985).

[22]   A.-C. Albertsson and O. Ljungquist, *J. Macromol. Sci. – Chem., A23*(3), 393 (1986).

[23]   M. Shimada and M. Takahashi, in *Wood and Cellulosic Chemistry* (D. N.-S. Hon and N. Shiraishi, Eds.), Dekker, New York, 1991, p. 621.

[24]   Professor Yoshiharu Doi, The Institute of Physical and Chemical Research, Japan, Paper Presented at the 2nd Bio/Environmentally Degradable Polymer Society Meeting, Chicago, Illinois, 1993.

SOURCE AND MAINTENANCE OF MICROORGANISMS USED FOR TESTING PLASTICS

JOAN KELLEY

International Mycological Institute
Bakeham Lane, Egham, Surrey TW20 9TY, UK

ABSTRACT

In some countries legislation on the use of biodegradable plastics has outpaced the development of reliable standard methods of measuring degradation. There are numerous methods designed to assess soluble and water-miscible products, but standards designed specifically for plastic materials are few. This paper discusses some current techniques and those under development with particular reference to the organisms employed. The need to use defined test strains, mixed organisms from the environment, or both is discussed along with the problems that can arise when attempting to maintain the relevant enzyme activities in test strains in the laboratory. The speed with which some isolates can lose activity and vigor during maintenance is assessed.

INTRODUCTION

Plastics are heterogeneous materials. They may contain impurities such as residual oligomers, monomers, reagents, and products of side reactions. They may also contain compounds included intentionally such as plasticizers, fillers, pigments, mold release agents, and starch added to enhance degradability. These must be borne in mind when selecting suitable organisms or sources of inocula to test their resistance or susceptibility to attack.

The eventual use and/or likely method of disposal is also of significance. The UK disposed of 85% of its domestic waste by landfill in 1990 while Denmark, for example, disposed of only 10% by this method [1]. This means that the bulk of the

UK plastic waste would experience anaerobic conditions with high levels of bacterial activity, while in countries where composting is of more significance, mixed populations of aerobic microorganisms will be the major degraders. Plastic waste which never finds its way into a disposal system and finishes its life as litter in soil contact will be heavily reliant on fungal species to bring about biodegradation as will biodegradable plastic mulches and other agricultural materials.

Wastes in fresh water or marine environments are more likely to encounter aerobic bacterial activity. Susceptible materials which have been stored badly at high humidity are likely to be attacked by fungi while those in water contact may be damaged by bacterial activity. All this must be taken into account when designing test regimes.

DEFINITIONS

Biodeterioration and biodegradation are terms which are often considered to be interchangeable. However, Heuck [2] defined biodeterioration as "any undesirable change in the properties of a material of economic importance caused by the vital activities of organisms." Biodegradation, on the other hand, has been defined as "the harnessing by man of the decay abilities of organisms to render waste material more useful or acceptable" [3]. This definition is inadequate for current requirements as it implies "the hand of man" must be present, but materials in a natural uncontrolled environment will obviously biodegrade. It does, however, indicate the desirable nature of biodegradation as opposed to the undesirable biodeterioration. An acceptable definition of the term "biodegradable" is yet to be agreed upon internationally. Is complete mineralization and removal from the environment necessary before a material is considered to be "biodegradable," or is breakdown to a smaller molecule acceptable? If so, how simple must the molecule be, must it be nontoxic (certainly) and biologically inert (in which case it is not biodegradable), and so the argument continues.

A number of bodies are working toward acceptable definitions, and there are drafts in various stages of preparation and completion. All extremes are covered with DIN requiring "naturally occurring metabolic endproducts" while the Biodegradable Plastics Society (BPS) in Japan suggests a biodegradable polymer is one capable of "being decomposed into low molecular weight components."

TESTS FOR BIODETERIORATION AND BIODEGRADATION

"A testing protocol should include an environment that fairly represents that to which the substrate polymer will be exposed" [4]. The above statement holds true when we know exactly what this final environment will be, and such factors must be taken into account when designing a test method. However, problems can arise on the numerous occasions when the end use/disposal is not known or the material may be destined for multiple fates as is more often the case. Simulated environment tests, however, only form one part of the total testing protocol which is required.

There have always been conflicting opinions regarding the source of inocula for testing. Many workers feel that only organisms freshly isolated from relevant substrates should be employed or that inocula based on soil or sewage with or without acclimatization are the sole answer. Many of the standard biodeterioration tests require the use of defined species (but not always strains).

There is justification for all points of view. In a research and development environment where specific conditions of use and disposal are known, then every effort should be made to obtain the relevant, active organisms that are likely to be encountered, either in isolation or as an acclimatized inoculum. However, rapid screening methods are often required in the early stages of a development program when the use of a standard test set can be helpful. The use of defined organisms to assess degradability or otherwise of additives (plasticizers, etc.) is also recommended. Standard sets of organisms are also useful to provide final "bench mark" testing which aids comparisons between products. Lee et al. [5] used pure fungal culture systems to allow the distinction between chemical and biological degradation of novel materials. Yakabe et al. [6] concluded that although the biodegradability of plastics in simulated natural environments is desirable, the reproducibility of these methods is not high, while the use of specific enzymes [7] or microorganisms [7, 8] can estimate biodegradability quickly and reproducibly.

Biodeterioration Testing

Table 1 shows some of the standard tests which are available for use in biodeterioration and biodegradation testing of plastics. The problems arising out of current biodeterioration tests and the organisms employed have been discussed elsewhere [9]. These standards are mentioned here because they contain techniques which are valid for both biodeterioration and biodegradation testing, e.g., soil burial methods. A number of workers have looked at the role of soil burial methods in biodegradation testing. Yakabe et al. [6] studied the factors affecting the biodegradability of polyester in soil.

Biodegradation Testing

The definition and measurement of biodegradation is hampered by the lack of recognized standards. There are three major questions which require an answer to predict the likely behavior of a plastic in the environment.

TABLE 1.   Some Standard Tests Designed for Use on Plastics

ISO 846	Plastics—determination of behavior under the action of fungi and bacteria
ASTM G21-90	Standard practice for determining resistance of synthetic polymeric materials to fungi
ASTM D 5209-92	Standard test method for determining the aerobic biodegradation of plastic material in the presence of municipal sewage sludge
ASTM D 5210-92	Standard test method for determining the anaerobic biodegradation of plastic materials in the presence of municipal sewage sludge

1. Is the material readily biodegradable? For this a simple stringent test is often used employing organisms and inocula with limited time for acclimatization.
2. Is the material inherently biodegradable? Current tests usually provide more favourable conditions using larger inocula and acclimatized organisms. Limits may be set on all tests, for example, 60% degradation after 28 days may be considered to indicate biodegradability in the Sturm test.
3. Is the material going to degrade under field and waste disposal conditions? The tests here tend to be simulated environment techniques.

The terminology and early biodegradability tests were mainly devised to cover liquid and sparingly-soluble or limited water-immiscible liquids such as detergents, surfactants, oils, etc. Testing of water-soluble polymers is consequently relative easy; methods include biological oxygen demand (BOD), carbon dioxide evolution (e.g., Sturm test [10]) and semicontinuous activated sludge tests (SCAS) [11]. Water-insoluble plastics, however, are more difficult. Two relatively new ASTM tests [12, 13] (Table 1) have begun to address the problem. Degradation of a known standard is employed as an organism control and 70% degradation of this material is required to confirm a valid test. These, in a program together with the established ASTM Biodeterioration tests [14–16], can give the experimenter a great deal of information. The International Biodeterioration Research Group (IBRG) is currently ring testing a modified Sturm test for use with biodegradable plastics. There are a number of techniques employed to simulate environmental situations, e.g., soil burial [16], simulated landfill, and simulated composts. Biodegradation may be assessed by weight loss, gas evolution, etc.

Many naturally occurring polymers can take a number of years to degrade, e.g., some timbers, pine needles. There are examples of landfill sites where newspapers are still legible after 40 years; this supposedly readily degradable polymer did not degrade in this field situation. We must ensure that we are not demanding more from biodegradable plastics than we expect from naturally occurring polymers.

EXPERIMENTAL WORK

There is a case for the use of defined cultures within the testing protocols of plastics materials. However, if test organisms are to be employed, it becomes imperative that the correct strains are selected and that these strains are well maintained. The International Mycological Institute (IMI) has had a rolling program of monitoring fungal test strains and screening new isolates with a view to improving test methods. This work has fallen into three stages: investigations of the variation in enzyme activities within species, looking at the variation between test strains held in different culture collections, and assessing the effect of maintenance regimes on enzyme activities. Data resulting from the first two studies have been published [17]. The work reported here completes one aspect of part three above.

METHODS

A series of semiquantitative screening tests has been developed to investigate enzyme activities in fungal cultures and has been described previously [17]. These methods were used to investigate the effect of subculturing strains as a means of culture maintenance.

There are a number of recognized test strains held by the IMI Genetic Resource Collection which are cited in standard test methods. Eighteen of these strains (Table 2) were put through six of the enzyme screens to investigate amylase, cellulase, protease, polycaprolactone degradation (as an indicator of potential to degrade polyurethanes, plasticizers, etc.), lipase, and pectinase activities. The tests were selected to monitor the activities of enzymes which could be of significance in the breakdown of commercial plastic formulations and plastics designed or amended to be biodegradable.

All results were rated on a 0–5 scale from zero to very high activity. Growth was also assessed and designated A–E, indicating very good growth to no growth, so that A5 would be the rating of a strain giving good growth and high enzyme activity.

The 18 isolates were then subcultured weekly for 24 weeks onto their recommended growth media, i.e., no special precautions were taken to ensure enzyme activities were retained by challenging with the relevant inducing substrate. The subcultures were grown for 7 days and then put through the screens described above. After 24 weeks the cultures were subbed onto media containing relevant inducing substrates for a further 10 weeks. Once more the cultures were screened after each subculture to assess recovery of enzyme activities.

RESULTS

Table 3 summarizes the overall results after 24 weeks subculturing. A number of activities were very rapidly lost or much reduced. *Chaetomium globosum* 16203 lost enzyme activities allowing polycaprolactone degradation after one subculture; *Stachybotrys atra* 82021 lost this ability after two subcultures. Other strains lost this activity after five and six subcultures. Amylase activity, important in the degradation of starch-based and amended plastics, was lost by *Scopulariopsis brevicaulis*

TABLE 2.   Organisms Tested

Fungal species	IMI strain numbers
Aspergillus amstelodami	17455
A. flavus	91856
A. niger	17454 and 91855
A. terreus	45543
A. versicolor	45554
Aureobasidium pullulans	45533
Chaetomium globosum	45550 and 16203
Paecilomyces vaiotii	108007
Penicillium cyclopium	19759
P. funiculosum	14933, 211742, and 87160
P. ochrochloron	61271
Scopulariopsis brevicaulis	49528
Stachybotrys atra	82021
Trichoderma viride	45553

TABLE 3.   Enzyme Activity Levels in Fungal Test Strains after Twenty-four
Subcultures

Enzyme activity	Activity level maintained	Activity level reduced	Activity level not detectable	No initial activity
Amylase	5	5	7	1
Cellulase	5	7	3	3
Protease	2	4	8	4
Polycaprolactone degradation	6	4	8	0
Lipase	0	3	13	2
Pectinase	4	12	2	0

49528 after four subcultures and by *Stachybotrys atra* 82021 after seven subcultures.
Cellulase activity in two *Chaetomium globosum* strains (45550 and 16203) was
reduced from Level 5 to Level 1 after six and seven subcultures respectively.

In general, lipase activity seemed to be the most readily lost with no strains
retaining maximum activity, 13 losing it completely. The ability to break down
polycaprolactone was retained by the highest number of strains but also had the
second largest number of total losses together with protease activity. Only two
strains lost pectinase activity completely but only four retained full activity and 12
showed much reduced activity. The ability to degrade polycaprolactone and pectinase activity was found in all organisms tested.

After 24 weeks, 10 of the 18 strains still showed some ability to degrade
polycaprolactone (six fully maintained) and again 10 strains still showed amylase
activity (five fully maintained).

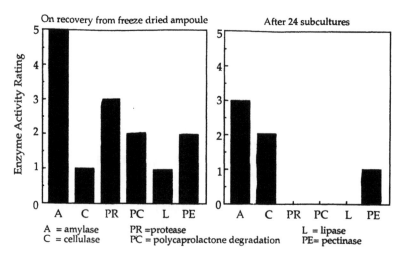

FIG. 1.   Enzyme activity ratings before and after subculturing *P. cyclopium* 19759.

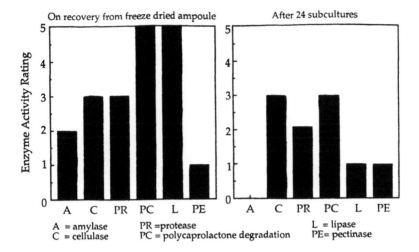

FIG. 2.   Enzyme activity ratings before and after subculturing *A. niger* 91855.

Figures 1 and 2 show profiles of initial activity and that after 24 weeks for *Penicillium cyclopium* 19759 which had 50% of the activities tested no longer detectable and *Aspergillus niger* 91855 which retained all but amylase activity.

Table 4 summarizes the results at 34 weeks after transfer to inducing substrates. Amylase and pectinase activities were not regained by any strains which had lost detectable activities. Three strains regained cellulase and protease activities and one regained lipase. Many strains showed further reductions in activity, and a number lost detectable activities completely.

DISCUSSION

The results from part one of these studies [17] suggested that when developing test methods it is not sufficient to simply recommend species. The variations in activities between strains (Fig. 3) indicated that optimal activities must be screened

TABLE 4.   Responses of Tests Strains after Transfer to Inducing Substrates between Twenty-four and Thirty-four Subcultures

Enzyme activity	Activity fully recovered	Activity maintained at reduced level	Activity further reduced	Activity lost
Amylase[a]	0	14	2	5
Cellulase[b]	3	7	1	3
Protease[b]	3	5	2	2
Lipase[b]	1	2	0	2
Pectinase[b]	0	9	4	0

[a]Extra strains were added to investigate amylase activity.
[b]Not all strains continued to be studied in this part of the work.

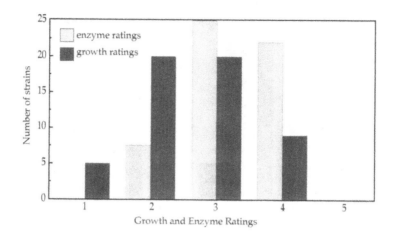

FIG. 3. Polycaprolactone degradation by *A. niger*. Reprinted from *International Biodeterioration, 24*, 289 (1988) with kind permission from Elsevier Science Ltd., The Boulevard, Langford Lane, Kidlington OX5 1GB, UK.

for, the best strains selected, and then referred to by culture collection number in the standard. Part two of the work showed that when cultures are well maintained by freeze-drying or in liquid nitrogen, optimal activities can be successfully retained as demonstrated by testing the same cultures held at IMI and at Centraalbureau voor Schimmelcultures (CBS), Baarn, the Netherlands.

The work described here demonstrates the importance of good maintenance techniques. While the tendency of organisms to become "laboratory strains" if maintained by subculturing is well recognized, the speed with which this occurred in some of these strains was surprising. It could be predicted that some strains having inducible activities would lose and then regain detectable activities when presented with the correct substrate. Some constitutive activity was retained throughout, but concern arises from the number of strains which lost activities rapidly and never regained them. This was presumably due to selection during subculture of conidia without the genetic information coding for these activities, and in this case the potential for change is probably even greater in fungi than in bacteria. Changes may also be due to heterokaryosis and the parasexual life cycle of fungi and to well documented intracellular DNA damaging processes which occur at a fairly constant low rate [18]. Maintaining fungi in an actively metabolizing state will inevitably introduce more of these problems when compared to freeze-drying or liquid nitrogen storage.

Maintaining "fed" cultures in fermenters etc. is not always the answer either, because while maintaining enzyme activities the ability to grow well and sporulate normally at air/substrate interfaces may be reduced with the organisms becoming "reactor" strains.

CONCLUSIONS

Testing protocols for plastics should always include environmental simulation tests and exposure to multiple organism inocula (sewage, soil), but there is an argument for the use of defined test cultures of bacteria and fungi within the development program. When defined fungal cultures are employed, these should include test strains identified by culture collection number. The use of a defined "core" set should not preclude the inclusion of other additional organisms selected by the worker.

Test strains should not be maintained routinely by subculturing and should be monitored for required activities at regular intervals.

REFERENCES

[1]   R. Glover, *Int. Biodeterior. Biodegrad., 31*(3), 171 (1993).

[2]   H. J. Heuck, *Mater. Org., 1*(1), 5 (1965).

[3]   D. Allsopp and K. J. Seal, *Introduction to Biodeterioration*, Edward Arnold, London, 1986, p. 136.

[4]   G. Swift, in *Agricultural and Synthetic Polymers Biodegradability and Utilization* (J. E. Glass and G. Swift, Eds.), American Chemical Society, Washington, D.C., 1990, p. 2.

[5]   B. Lee, T. L. Pometto, A. Fratzke and T. B. Bailey, *Appl. Environ. Microbiol., 57*(3), 678 (1991).

[6]   Y. Yakabe, K. Nohara, T. Hara, and Y. Fujino, *Chemosphere, 25*(12), 1879 (1992).

[7]   J. Hosokaw, M. Nishiyama, K. Yoshihara, and T. Kubo, *Ind. Eng. Chem. Res., 29*, 800 (1990).

[8]   Y. Tokiwa, T. Ando, and T. Suzuki, *J. Ferment. Technol., 54*, 603 (1976).

[9]   J. Kelley and D. Allsopp, in *Industrial Microbiological Testing* (J. W. Hopton and E. C. Hill, Eds.), Blackwell Scientific Publications, Oxford, 1987, p. 23.

[10] R. N. Sturm, *J. Am. Oil. Chem. Soc., 50*, 159 (1973).

[11] ASTM Test Method D2667-89.

[12] ASTM Test Method D5209-92.

[13] ASTM Test Method D5210-92.

[14] ASTM Test Method G21-90.

[15] ASTM Test Method G22-76 (1990).

[16] DIN Standard 53739.

[17] J. Kelley and P. A. Yaghmaie, *Int. Biodeterior., 24*, 289 (1988).

[18] P. Strike and F. Osman, in *Stress Tolerance of Fungi* (D. H. Jennings, Ed.), Dekker, New York, 1993, p. 297.

A COMPARATIVE STUDY OF THE DEGRADABILITY AND RECYCLABILITY OF DIFFERENT CLASSES OF DEGRADABLE POLYETHYLENE

S. AL-MALAIKA,* S. CHOHAN, M. COKER, and G. SCOTT

Polymer Processing and Performance Group
Department of Chemical Engineering and Applied Chemistry
Aston University
Birmingham B4 7ET, England, UK

R. ARNAUD and P. DABIN

Laboratoire de Photochimie

A. FAUVE

Laboratoire de Chimie Organique et Biologique

J. LEMAIRE

Laboratoire de Photochimie

URA CNRS 433, Universite Blaise Pascal
Clermont Ferrand, France

ABSTRACT

Abiotic degradation of representative samples from available commercial photo(bio)degradable polyethylene films was examined with respect to the rate and extent of degradation, oxidation products, and changes in molecular weight both during outdoor exposure and in laboratory photoaging devices with different accelerating factors. Although the rate of photoxidation was found to depend on the type of degradation system used, all the samples showed rapid rate of carbonyl formation with concomitant reduction in molecular weight and mechanical properties on exposure to UV light. The photofragmented polymers were shown to be much more hydrophilic in nature compared to the unoxidized analogues, and photofragments of all samples were found to contain high levels of low molecular weight (bioassimilable) carboxylic acids and esters. Recycling behavior of virgin polyolefins, both as homopolymers and heterogenous polymer blends, which contained 10% of nonoxidized and photooxidized photo(bio)degradable plastics was examined. It was found that the initial mechanical performance of homogeneous blends was not greatly affected by the presence of nonoxidized degradable materials. However, blends containing degradable films which were initially partially photooxidized had a much more detrimental effect on the recycled blends properties during processing and weathering; the effect was minimal for degradable polymers containing the iron–nickel dithiocarbamate system.

INTRODUCTION

All commercial polyolefins will degrade eventually as they are not chemically "pure" materials: they contain several chemical impurities which are introduced at different stages of their life cycle (polymerization, processing, manufacturing, storage, and upon environmental exposure in-service) [1]. Impurities such as metal ions, peroxides, unsaturation, and carbonyl compounds initiate oxidation of the polymer, and even stabilized polymers undergo eventual oxidation with the introduction of a variety of carbonyl species, in particular carboxylic acids and esters [2]. Antioxidants contribute to prolonging the product outdoor lifetime; this aspect raises many environmental problems in the form of plastics litter and as hazards to animals. Plastics products in many applications, e.g., high volume packaging and agricultural, are often made to be too stable for their intended purpose. In response to these problems, manufacturers produced a number of polyolefin-based materials with enhanced degradability, and products based on these materials have been claimed to be photodegradable, biodegradable, or both, often without defining the conditions or the way in which the polymer product degrades. The lack of sufficient scientific evidence and objective criterion to test the manufacturer's claims on the performance and degradability attributes of their products has raised public concern [3]. We report here some of the work carried out under the Brite-Euram collaborative research program [4] which was set up to examine how chemical, physical, and biological factors affect the degradability of commercially available photo(bio)degradable polyethylenes. Some of the issues addressed in this research program have already been reported [5]. Three classes of commercial degradable polyethylene-based materials were used throughout this work:

A.   Photolytic polymers (A samples): In this class the photosensitizing group forms an integral part of the polymer chain. Two different types of polyethylene (PE) containing carbonyl groups are available commercially:

(i)   The carbonyl is built in the polymer chain by copolymerizing carbon monoxide with ethylene, ECO system [6]; this is now used for specialized applications such as "six-collar packs" for beverage cans.

(ii) Polyethylene containing copolymerized vinyl ketone in the side chain, Ecolyte (Guillet process) [7].

B. Polyethylene containing metal compounds as photoactivators (B samples):

(i)   Polymer containing organo-soluble transition metal ion (generally iron in the form of carboxylate [8], e.g., acetyl acetonate, FeAcAc) [8a], and

(ii) Polyethylene containing metal chelates (with sulfur ligands) which act as antioxidants and photoactivators (AOPA); e.g., Fe^{III} dithiocarbamate, FeDRC (Scott–Gilead process) [9].

C. Starch-filled polyethylene (C samples): PE-starch formulation (Griffin process) [10, 11] may be subdivided into two categories:

(i)   Polymer containing starch, normally at 6–8%, and

(ii) Polymer containing starch and iron carboxylate (generally stearate, FeSt).

FeAcAc   FeDRC   FeSt

This paper reports on some aspects of abiotic degradation and recycling potential of representative samples of each of the above materials, together with degradable systems, "similar" to the commercial Scott–Gilead polymers, prepared in the laboratory under controlled and known thermal history. The key role of the prior oxidation on the subsequent biodegradation of these polymers will be highlighted but their biodegradative behavior, which has also been examined [4], will not be discussed.

EXPERIMENTAL

Materials and Recycling Operation

Tables 1a and 1b show the origin and composition of representative samples of commercially available photo(bio)degradable polyethylene films (thickness: 12–40 μm except **A3** which was 508 μm). All the samples were low density polyethylene (LDPE)-based except for **B7** and **B8** which were high density polyethylene (HDPE). Virgin unstabilized polymers, LDPE (Novex L-61, from B.P), PP (propathene HF23, from ICI) and EPDM (Vistalon 250, from Exxon) were used for recycling experiments. All solvents used were either HPLC or spectroscopic grades.

LDPE samples containing different concentrations of FeDRC (similar to the Scott–Gilead commercial degradable polymer) were extruded into thin films (75 μm) under controlled laboratory conditions using a single screw extruder, SSE (Humboldt HE 45-25DE, L/D ratio 25:1) at 45 rpm with the three barrel zones set

TABLE 1a. Commercial Degradable Low Density Polyethylene Samples (except B7 and B8 which are high density polyethylene). All Samples Are Thin Films Except A3 Which is a Thicker Sample from a "6-Collar Pack"

Type	Code	Origin	Appearance	Composition
Photolytic copolymer	A1	Ecolyte	Transparent	Ethylene–vinyl ketone copolymer
	A2	Dupont	Transparent	Ethylene–CO copolymer
	A3	ITW Hi-Cone PE	6-Collar pack	Ethylene–CO copolymer
Metal compounds	B1	Atochem degralence	Transparent	Fe stearate
	B01	Plastopil	Transparent, light gray	FeDRC
	B5	Plastopil	Transparent, light green	FeDRC/NiDRC
	B7	Enichem 221 HT	Transparent	FeAcAc/HDPE
	B8	Enichem 2HT	Transparent	FeDRC/HDPE
Starch-filled PE	C2	Ampacet	White	Starch + FeSt + TiO_2
	C3	Ampacet	White	Starch + FeSt + 3% TiO_2
	C4	Ampacet	Green	Starch + organic dye
	C5	Ampacet	Black	FeSte + carbon black

at 160°C and the die head temperature set at 180°C. Multiple processing (recycling) was carried out either in a closed chamber of an internal mixer (Rapra-Hampden torquerheometer, TR) for 10 minutes at 180°C and 60 rpm (thin films for subsequent analysis were compression molded at 180°C), or extruded in the SSE as above. Polymer blends containing 10% of either a nonoxidized or photooxidized degradable polymer were processed (as above) and a portion of the processed polymer was removed for testing (melt flow index, MFI, carbonyl index, molecular weight, and weathering) before reprocessing the remainder (second pass). This

TABLE 1b. LDPE Degradable Films (75μm) Extruded (SSE) in the Laboratory (at Aston Univ)

Code	Origin	Appearance	Composition
Bs0	Laboratory prepared	Transparent	LDPE, unstabilized
Bs01	Laboratory prepared	Transparent, light gray	FeDNC, 0.01%
Bs02	Laboratory prepared	Transparent, light gray	FeDNC, 0.02%
Bs03	Laboratory prepared	Transparent, light gray	FeDNC, 0.03%
Bs04	Laboratory prepared	Transparent, light gray	FeDNC, 0.05%
Bs3	Laboratory prepared	Transparent, light green	FeDNC/NiDNC, 0.03%/0.01%

process was repeated twice more with further sampling and testing after each pass. Five polymer blend series, each containing 10% degradable polymers, were examined; LDPE (series I and II, with nonoxidized and photooxidized degradables, respectively); PE:PP at a ratio of 1:1 (series III and IV with nonoxidized and photooxidized degradables, respectively); and PE:PP:EPDM at a ratio of 1:1:0.2 (series V with nonoxidized degradables).

UV Exposure and Monitoring of Change in Properties and Degradation Products

Exposure of polymer samples was carried out under natural conditions (outdoor exposure in Birmingham facing south at 45°) and laboratory conditions using two different accelerated photoaging devices. Photoaging was performed in the mildly accelerated environment of a sunlamp-blacklamp (S/B) cabinet, comprising 7 sunlamps and 21 actinic blue lamps with a maximum relative intensity within the 280–370 nm range (radiation output of about 4.5 $W \cdot h \cdot m^{-2}$) and an average temperature of 39°C, and under a highly accelerated photoaging environment of a Sepap 12/24 unit (at 60°C) fitted with four 400 W medium pressure mercury lamps with the radiation filtered at $\lambda > 300$ nm. The rate of photooxidation of thin films was determined by monitoring changes in the carbonyl and, in some cases, the double bond area index using Fourier transform infrared analysis (FT-IR) which was run on a Perkin-Elmer PE-1710 FTIR and PE-data manager program (area index calculated by the PE-data manager program using predefined baseline for the functional group absorption peak over that of a reference peak due to absorption of a $-CH_2-$ in the backbone of the polymer which does not change during oxidation). For the laboratory-extruded PE samples containing a different initial [FeDRC], the concentration of photooxidation products (acid, ester, and ketone) was calculated after derivatization with SF_4 according to a literature method [12, 13]. Melt flow index (MFI) was measured on a Davenport polyethylene grader at 190°C (0.118 cm die, 2.16 kg). Molecular weight was determined by gel permeation chromatography using a Waters 150C/A LC chromatograph and results are expressed in equivalent weight of polystyrene. Mechanical properties were determined from tensometric measurements (using a micro 500 Testometric tensometer according to ASTM standard, D882. Extraction of photofragmented degradable polymers (degraded beyond embrittlement) was carried out with either an aqueous alkaline solution of different strengths (0.001 to 1.0 M NaOH) or with water. The extracted polymer fragments and the alkaline aqueous extracts, after neutralization and extraction with ether (in the case of water extracts, no further treatment was carried out except for concentrating the solution), were characterized by spectroscopic methods (FT-IR, using KBr disk and FT-NMR).

RESULTS AND DISCUSSION

Photooxidation

The rate and extent of photooxidation of representative samples of degradable polymers were studied by following the reduction in molecular weight and intrinsic viscosity (by GPC) and the build-up of carbonyl groups and, in some cases, unsaturation (by FT-IR). All the commercial degradable samples exhibited a rapid rate of carbonyl formation and molecular weight reduction on exposure to UV light. Figure 1 shows typical changes in functional groups (carbonyl and unsaturation) concentration during UV exposure (in S/B cabinet) of degradable films, e.g., for BsO3. The formation of the different carbonyl groups in LDPE, viz., carboxylic acid (1715 cm^{-1}), ketones (1720 cm^{-1}) and esters, both acyclic (1739 cm^{-1}) and cyclic (1785 cm^{-1}), and the vinyl (909 cm^{-1}) group occur by known mechanisms of photooxidation (Ref. 14 and references therein), see Scheme 1.

Transition metal ions, e.g., iron and cobalt, are known to be effective initiators for photooxidation since they catalyze the rate of hydroperoxide formation via the redox reactions 1 and 2, hence a number of commercial polymers designed for degradability contain iron ions as effective photoactivators. Figure 2 compares the rate of photooxidation (exposure in S/B cabinet) of representative samples of commercial PE-based degradable films containing transition metal ions.

$$Fe^{2+} + ROOH \longrightarrow Fe^{3+} + RO\cdot + OH^{-} \qquad (1)$$
$$Fe^{3+} + ROOH \longrightarrow Fe^{2+} + ROO\cdot + H^{+} \qquad (2)$$

All polymer samples containing iron compounds (FeAcAc, B7, FeDRC, B8, B01, and the starch-filled FeSt containing polymer, C2) photooxidize without an induction period whereas the nickel-iron dithiocarbamate, Ni/Fe, containing polymer B5, shows a well-defined induction period followed by rapid photooxidation. The rapid increase in carbonyl concentration during UV irradiation, which is paralleled by reduction in molecular weight and intrinsic viscosity of the polymer (see inset Fig. 2b), is due to photoreduction of the trivalent iron complex, e.g., iron carboxylate, with the formation of initiating radicals leading to oxidative breakdown of the polymer, Reaction (3) [8]:

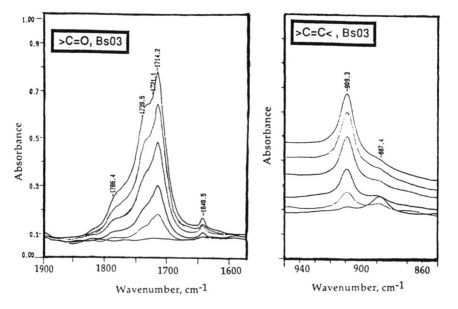

FIG. 1.   Development of carbonyl and unsaturation groups (from IR) during photo-oxidation of laboratory-prepared FeDRC (0.03%) containing PE film.

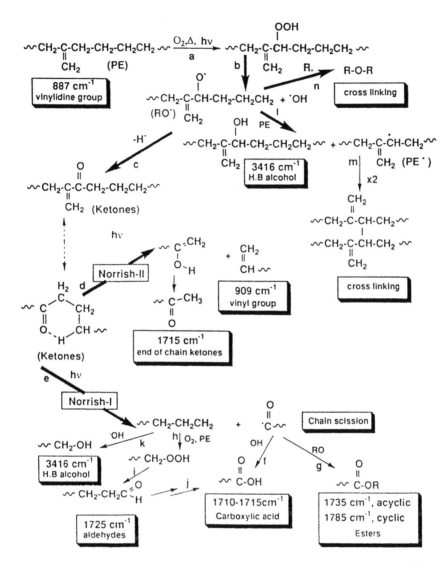

SCHEME 1.   Oxidation of LDPE during processing and exposure to UV light. Observed IR frequencies are shown; see Figs. 6 and 7.

$$(RCOO)_3Fe \xrightarrow{h\nu} (RCOO)_2Fe + RCOO$$
$$\downarrow$$
$$CO_2 + R \qquad (3)$$

FeDRC-containing polymer (B8), inset Fig. 2(a). While FeSt, FeAcAc, and FeDRC are effective photosensitizers, the first two, unlike FeDRC [8a, 15], are not thermal (processing) antioxidants. Peroxidolytic metal dithiolates (e.g., dithiocarbamates, dithiophosphates, xanthates) are good processing antioxidants and many, particularly Ni and Co complexes, are also very effective photoantioxidants [16]. Iron complexes, e.g., FeDRC, however, are less photolytically stable than their Ni ana-

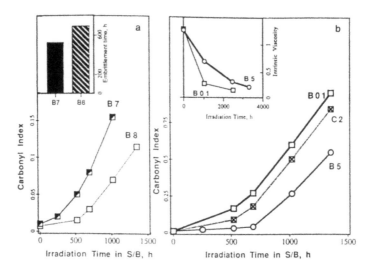

FIG. 2. Rate of photooxidation (S/B cabinet) of Fe-containing commercial PE degradables. Inset (a) shows embrittlement time for B7 and B8, and inset (b) shows changes in the intrinsic viscosity (from GPC) of Samples B01 and B5 during UV exposure.

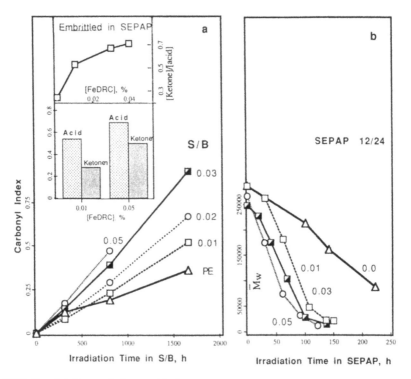

FIG. 3. (a) Rate of photooxidation (S/B cabinet) of laboratory-extruded PE films containing FeDRC at concentrations shown on curves. Inset shows concentration of carboxylic acids and ketones formed in these samples on embrittlement (exposed in SEPAP 12-24 device) after SF_4 derivatization. (b) Rate of reduction in $\overline{M}_w$ of these samples during UV exposure.

logues; FeDRC photolyzes by Reaction (4) in the absence (or presence, if used at high concentration) of an induction period, IP [15]. It was shown previously [15] that the embrittlement time for FeDRC-containing PE films is concentration-dependent. Laboratory-extruded LDPE films containing different low concentrations of FeDRC (in the range of concentration used in the Scott–Gilead process) showed higher rates of photooxidation and reduction in molecular weight at higher [FeDRC], see Fig. 3. This finding is further supported by the formation of higher concentrations of carboxylic acids and ketones in the photoembrittled samples containing higher initial [FeDRC], see inset Fig. 3(a) (calculated following SF_4 derivatization [13]). While the concentration of carboxylic acids formed at all FeDRC concentrations was higher than that of esters and ketones, the ratio of ketone to acid increased with higher FeDRC concentration, see inset Fig. 3(a). It is clear from these results that the major final photooxidation product in FeDRC-containing polymers is carboxylic acids and, to a lesser extent, esters, formed via Norrish I photolysis of intermediate ketones that are, in turn, formed from thermolysis (during processing) and photolysis of polymer hydroperoxide, see Scheme 1. Norrish II photolysis, which gives rise to end-chain ketones and vinyl groups (see formation of vinyl absorption peak at 909 cm^{-1} in Fig. 1), appears to play a more important role when higher FeDRC concentrations are initially added to the polymer. Both Norrish I and II processes lead to chain scission, which is reflected in the observed reduction in molecular weight at all [FeDRC], see Fig. 3(b), but only Norrish I leads to the formation of free radicals [17].

At much higher FeDRC concentration an IP becomes apparent [15, 18]. The length of this IP, however, can be controlled more effectively, while maintaining the subsequent fast rate of photooxidation, by using low concentrations of FeDRC in combination with very small concentration of the more photolytically stable nickel dithiocarbamate, NiDRC [9, 18]. The commercial sample B5, which is based on a Ni/Fe dithiocarbamate system, illustrates this clearly (Fig. 2b). Like other peroxide decomposers, the antioxidant effect of the dithiocarbamates is due to oxidation products of the thiocarbomoyl moiety, Reaction 4b, giving low molecular weight sulfur acids which are responsible for the nonradical decomposition of hydroperoxides [19]. Once the iron complex is destroyed by light, however, the free iron ion released (in the form of macromolecular carboxylate, Reaction 4a) undergoes the normal redox reaction with hydroperoxides (see Reactions 1 and 2), leading to the rapid photooxidation and reduction in molecular weight of the polymer, see inset Fig. 2(b). It is this dual activity of the FeDRC, as an antioxidant and photoactivator, which has led to the development of the Scott–Gilead process for degradables [9].

Exposure of the different photo(bio)degradable polymers outdoor and in the accelerated photoaging cabinets (S/B and SEPAP) gave good agreement in terms of the relative rates of photooxidation of the different commercial samples; see, for example, Fig. 4(a). Samples of starch-filled polyethylene containing ferric stearate (C2 and C3, Fig. 4a and b) showed the highest rate of photooxidation (C2 photooxidizes at a rate similar to films containing FeDRC, B01) due to the photoactivation role of the iron ion. On the other hand, films containing starch and an organic dye (C4) only showed a much reduced rate of photooxidation, not too dissimilar from a control LDPE film. This observation suggests clearly that starch does not mediate in the photooxidation of the polymeric matrix. Further, starch was shown [4] to exert no influence on the subsequent bioassimilation process unless the polymer substrate has been extensively photooxidized (with much reduced molecular weight, e.g., M_w 2000) and the starch has been released from the polymer. On the other hand, an LDPE film containing iron stearate and carbon black, C5, showed a fast rate of photooxidation, Fig. 4. This behavior is unexpected since carbon black (a highly UV-absorbing additive) is expected to prevent the photoexcitation of iron stearate. The possibility of a synergistic effect involving both the oxidation of carbon black and polymer surfaces may explain this observation.

The photoxidation rates of the photolytic polymers, the "Guillet" Ecolyte and the ECO copolymers (A1 and A2, respectively), and the "Scott–Gilead" Ni/Fe dithiocarbamate polymer, B5, are compared in Fig. 5 (inset). In contrast to B5, the photolytic polymers are immediately photooxidized on exposure to light without an

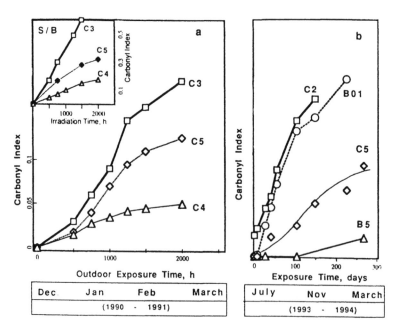

FIG. 4.   Rate of photooxidation of starch-containing polymers (C samples), FeSt-carbon black (C5) and Ni/FeDRC (B5) polymers during outdoor exposure (in Birmingham). Inset compares rates of photooxidation of starch-containing films exposed in S/B cabinet.

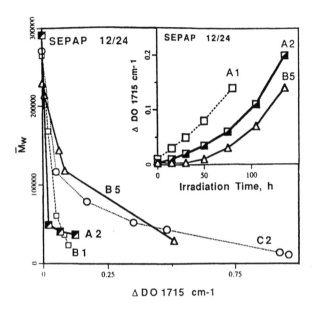

FIG. 5.   Rate of reduction in $\overline{M}_w$ of ECO (A2) and Fe-containing polymers (B1, B5, C2) during UV exposure (in SEPAP 12-24) as a function of change in carbonyl absorption.

IP. A comparison of the rate of reduction in molecular weight of the ECO polymer (A2) with iron-containing polymers (see Fig. 5) reveals that, in all cases, the rate of reduction in molecular weight is fast at the early part of the photooxidation and is particularly rapid in A2 (M_w generally decreased from about 300,000 to 50,000–20,000, with a further decrease to 6000 to 2000 on prolonged UV-exposure well beyond the embrittlement time of the polymer).

While the molecular weight of all other commercial systems examined continued to decrease gradually on prolonged exposure, the rate of reduction in the case of the ECO polymer, A2, at this stage was much slower. The behavior of A2 suggests that the degree of degradation in this case is strongly dependent on the initial carbonyl concentration in the degradable polymer. An important finding is that when the ECO (A2) and the FeDRC (B01) containing polymers were photofragmented (in the S/B cabinet well beyond embrittlement) to the same extent in terms of molecular weight reduction, to about 96% of their original value (MW decreases from 254,200 to 9,600 and from 240,600 to 5,800, respectively) and the total carbonyl content was compared in the fully fragmented samples, it was found that the ECO polymer, A2, had a lower overall carbonyl content compared to the FeDRC polymer; A2 contained only 63% of the total carbonyl content of the FeDRC-containing polymer sample, B01. The lower content of carbonyl products in the ECO polymer, A2, makes the fully fragmented polymer more hydrophobic than the B01 sample degraded to the same extent, and this can account for the observed [4] rapid microbial growth in the FeDRC-containing polymers compared to the ECO polymer which showed faster initial degradation (in MW reduction) but underwent much slower biodegradation.

Degradation Products of Photodegradable Polyethylene Films

It was shown earlier (see Fig. 1) that all the photodegraded polymers examined gave large amounts of carbonyl-containing compounds: primarily acids, esters (acyclic and cyclic), and ketones. Although these are generally known to be nontoxic to animals, there is public concern [3] about the lack of information on the possible hazards associated with the degradation products of degradable plastics. To address the question of whether the oxidation products of photooxidized and fragmented PE-based degradables could be removed from the polymer under environmental leaching conditions, samples B5 (Ni/FeDRC system) and B7 (FeAcAc) were exposed to the sun and rain, and the liquid washing the surface of the polymer samples was removed at intervals and examined for residues after evaporation; no aqueous extraction was observed from either sample. Furthermore, it was found that even for very heavily photooxidized samples (in S/B cabinet well beyond embrittlement), the oxidation products could not be extracted with water alone, nor was it possible to extract any significant amount of degradation products in the aqueous layer after dilute alkaline (0.01–0.1 M NaOH) treatment. However, the photofragmented (high carbonyl content) films did swell in water and could be readily suspended in a dilute alkaline solution due to their modified highly hydrophilic nature. It was concluded that the carboxylic acids and ester groups formed in the surface layers of the polymer are substantially aliphatic in nature with long alkyl chains and hence resistant to aqueous leaching. However, although normal groundwater cannot remove these low molecular weight oxidation products, it was shown [4] that microorganisms can remove these oxidation products which are located in or near the surface of the photofragmented films (they do so even in a nutrient-starved medium).

The nature of the degradation products was therefore further examined by modifying the extraction treatment; more concentrated (1 M) aqueous alkaline solution was found to remove the low molecular weight degradation compounds. Figure 6 shows the functional groups region (carbonyl and unsaturation, from FT-IR and solid-state ^{13}C NMR) of the ECO polymer A3 before exposure (A3-A), after photofragmentation in S/B cabinet (beyond embrittlement at a point where 96% reduction in MW has taken place, A3-B, and after photofragmentation and aqueous alkali extraction, A3-E). Photoxidation (A3-B) leads to the formation of acids, esters, γ-lactone, and ketones. The alkaline extraction leads to sharp reduction in the carbonyl absorption and the appearance of a carboxylate peak at 1574 cm^{-1}, following the conversion of acids (1715 cm^{-1}) and esters (1739 cm^{-1}) upon treatment with NaOH; ketones and aldehydes are unaffected by this treatment, see A3-E. Solid-state ^{13}C-NMR spectra of samples of the photofragmented, A3-B, and photofragmented and alkaline treated, A3-E, Fig. 6, support this and show clearly a shift in the carbonyl peak (broad) from the 169–175 ppm region (in A3-B) characteristic of carboxylic acids and esters to 181 ppm (in A3-E) attributable to the carboxylate group. Further resonance signals at $\delta = 114$ and 139 ppm and 74 ppm appear in both samples; these are attributable to unsaturated carbons of the vinyl group formed via Norrish-type II photolysis of the ketone carbonyls in the ECO polymer, and to secondary alcohols which may be formed from hydrogen abstraction and further oxidation of the active methylene next to the vinyl group, see Scheme 2 and Fig. 6. These results support the important role played by the Norrish II reaction (compared to Norrish I) in ECO polymers [17].

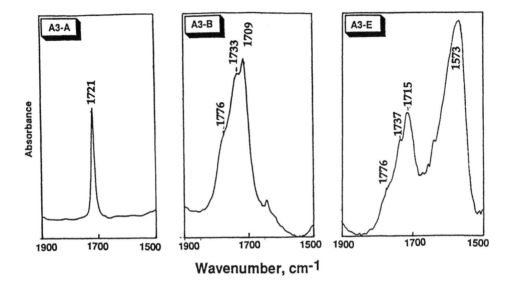

Wavenumber, cm⁻¹

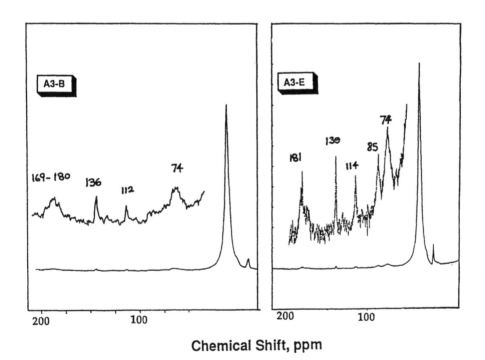

Chemical Shift, ppm

FIG. 6.   IR and solid-state ¹³C-NMR spectral regions for the ECO polymer A3 before UV exposure (A), after photofragmentation (B), and after photofragmentation and alkaline extraction (E).

Figure 7 shows FT-IR spectra in the carbonyl region of the photofragmented (in S/B cabinet, M_w decreased to 97% of its initial value) FeDRC-containing polymer film, B01, before B01-B and after B01-E with the same aqueous alkaline treatment, as well as that of the aqueous extract after neutralization and extraction in organic solvent, B01-D. Only about 10% of the total carbonyl products present in this photofragmented (M_w 5800) sample was found to be present in the aqueous alkaline extract; this is taken to suggest also the level of the extractable (in aqueous alkaline solution) low molecular weight oxidation products. FT-IR and ^{13}C and ^1H NMR (in carbonyl and hydroxyl regions) of extracted photooxidation products (after similar aqueous alkaline treatment) from photofragmented laboratory-prepared PE film containing a combination of FeDRC, 0.03%, and NiDRC, 0.01%, Bs3-D (i.e., "similar" to the commercial film B5) are also shown in Fig. 7. The extracted low molecular weight products in both cases (B01-D and Bs3-D) are shown to be composed mainly of carboxylic acid(s) (IR absorbance at 1712 cm^{-1} for B01 and at 1709 cm^{-1} for Bs3, as well as a ^{13}C-NMR signal at 180 ppm attributable to acid carbonyl) and to a lesser extent, alcohol (^1H signal at 8.5 ppm and a broad strong IR absorption peak at 3416 cm^{-1} attributable to strongly hydrogen-bonded OH group), see Fig. 7. The major contribution of the carboxylic acid to the final degradation product composition was confirmed from SF_4 treatment of photoembrittled laboratory-prepared PE-containing FeDRC samples, as shown earlier; see Fig. 3.

These results show clearly two major differences in the level of the oxidation products formed in the photofragmented samples of ECO (A3) and FeDRC (B01) polymers (degraded to the same level of MW reduction of about 96%), whereas the total carbonyl formed in A3 was (as shown earlier) about 60% of the amount formed in B01. The vinyl content in B01 was about 10% of that formed in A3. This finding clearly illustrates the greater contribution of the Norrish II photolysis in the ECO polymer, see Scheme 2, while photooxidation by Norrish I accounts for the formation of the final photodegradation products (mainly carboxylic acids) in the case of the commercial FeDRC-containing polymers, see Scheme 1. Furthermore, it can be concluded that the formation of these photooxidation products in all the degradable PE samples examined is of paramount importance as they not only lead to physical breakdown of the polymer, as evidenced by the loss of polymer strength, reduction in molecular weight, and photofragmentation, but also to making the originally hydrophobic polymer much more hydrophilic after degradation, hence accessible to the action of microorganisms during the biodegradation stage. The importance of the initial abiotic degradation and its role as a controlling step for subsequent biodegradation has been discussed recently [20].

Recyclability of Degradable Plastics

Recycling polymer products made from high value engineering plastics back to the original application, e.g., car bumpers to car bumpers [21], is a viable approach to the preservation of plastics; there is certainly no prospect for degradable materials in this form of recycling. However, low cost commodity plastics, e.g., polyolefins, are often recycled into products with inferior properties. A major problem of recycling in this case is the level of contaminants present, especially transition metal ions, which are potent catalytic degradants. To address the question

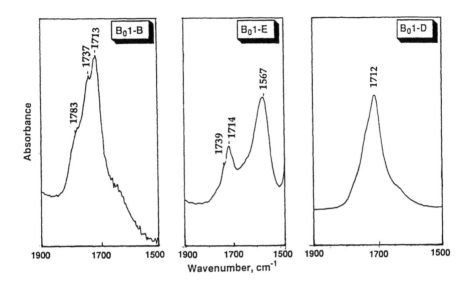

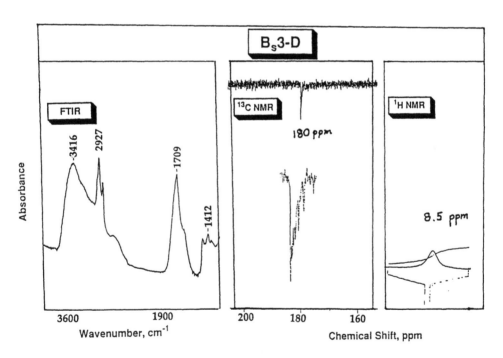

FIG. 7.   Carbonyl absorption in the IR spectra of the FeDRC-containing polymer B01 after photofragmentation (B), after photofragmentation and alkaline extraction (E), and for the aqueous liquid extract of the photofragmented sample (D). IR and specific regions of ^{13}C- and ^{1}H-NMR spectra of the liquid extract of a laboratory-extruded Ni/FeDRC-containing polymer treated in the same way (Bs3-D) are also shown.

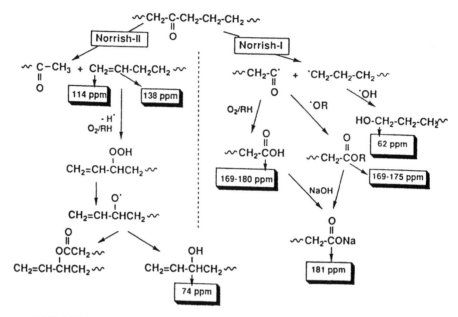

SCHEME 2. Photooxidation products of the ECO copolymers. The numbers are observed ^{13}C chemical shifts; see Fig. 6.

of whether homogeneous or heterogeneous (mixed) plastics waste can still be recycled if it was contaminated with small amounts of a degradable polymer, we have examined the recyclability of both homo- and heterogeneous polyolefin blends containing 10% of either nonoxidized or photooxidized degradables.

Compared to nonoxidized degradable polymer containing FeSt-starch filled, C2, and ECO copolymer, A3, the dithiocarbamate-containing polymers, B5 and B01, showed the least adverse effects on the recyclability of "homogeneous" blends with PE (see Fig. 8a) and "heterogenous" blends with PE:PP:EPDM (see Fig. 8b) since very little change was observed in the initial carbonyl index and melt stability (MFI) of the blends even after four processing passes. This is almost certainly due to the thermal antioxidant effect of the dithiocarbamate ligand [15, 16, 18, 19]. Multiple extrusion in SSE of nonoxidized degradables in PE, e.g., A3 and C2 (see Fig. 8c), has led to an increase in the molecular weight of the polymer (crosslinking) at lower temperatures (180°C), but this did not greatly affect the initial mechanical performance of the blends (see Fig. 8d). However, multiple extrusion of the same blends at higher temperature, e.g., 210°C, caused a progressive reduction in molecular weight (chain scission) together with a detrimental effect on the initial mechanical properties of the blends; controlling the processing parameters can therefore minimize the adverse effects of these degradables on the final properties of polymer blends. Figure 9 shows the effect of the presence of degradables which have been initially photooxidized on the reprocessibility of homogeneous (series II, PE) and heterogeneous (series IV, PE:PP) blends. The behavior of recycled blends containing the photooxidized Fe/NiDRC system, B5, is quite interesting since this shows that they can withstand the effect of reprocessing (up to four passes), whereas analogous blends containing FeSt-starch filled polymer, C2, and the ECO copoly-

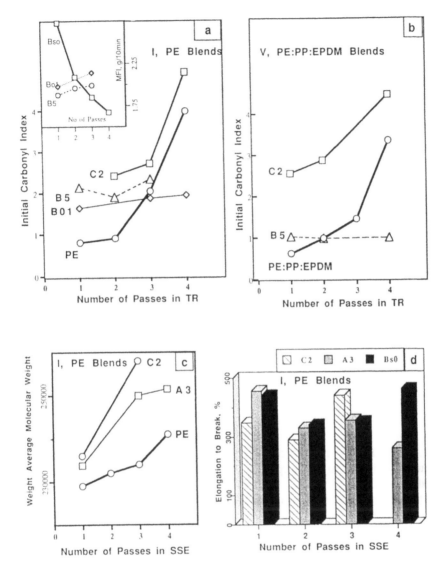

FIG. 8.   Effect of multiple processing in internal mixer (TR) and single screw extruder (SSE) at 180°C on melt stability (MFI and MW), carbonyl and elongation to break of blends containing 10% nonphotooxidized degradables.

mer, A3, showed severe oxidation (see Fig. 9). Multiple processing of blends containing degradables has generally led to a decrease in the vinylidine group (887 cm^{-1}) content with a concomitant increase in carbonyls and a decrease in MFI (i.e., increased molecular weight), but with relatively small change in the vinyl (909 cm^{-1}) concentration, especially during the first two or three passes; see, for example, Fig. 9a inset for blends containing photooxidized C2. This confirms [22] the importance of vinylidine hydroperoxide thermolysis and subsequent crosslinking reactions (decrease in MFI) of the derived alkoxyl or alkyl radicals, see Scheme 1

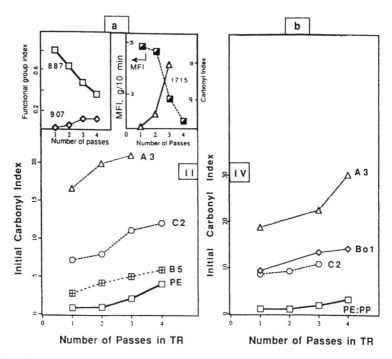

FIG. 9.   Effect of multiple processing in internal mixer (TR) at 180°C on melt stability (MFI) and changes in functional group concentrations of blends (Series II PE and Series
IV PE:PP) containing 10% photooxidized degradables, except for C2-III in Inset b, which
contains nonphotooxidized C2.

(a, b, l, m, n). The thermolysis of the ketones formed (Scheme 1c) via a Norrish II
reaction gives rise to vinyl groups and does not seem to play a major role during
processing.

The photooxidative stability of PE and PE:PP blends containing initially
photooxidized degradables is shown in Fig. 10. Again the Ni/FeDRC-containing
polymer (B5) was least detrimental to the "weathering" behavior of blends containing it, even though this was an initially photooxidized B5 sample, Fig. 10a. However, in general, all blends containing photooxidized degradables showed inferior
performance compared to the nonoxidized degradable analogues, both during processing and subsequently on exposure to UV light (see, for example, Fig. 10c and d
for sample C2); sample B5-II (photooxidized B5 in PE) which contains Ni/FeDRC
was least detrimental to the photostability of the PE blend. Furthermore, processing
severity increases the extent of photooxidation of these blends (inset Fig. 10a). It is
interesting to note, however, that compared to blends containing photooxidized
degradables, those containing nonphotooxidized degradables showed very little adverse effect on their UV stability (i.e., when compared to the stability of the polymer control), see Fig. 10c (curves PE and PE-containing nonphotooxidized and
oxidized C2). However, the final photooxidative stability of plastics containing
degradables depends on the oxidizability of the virgin polymer(s) used. For example, PE:PP blends are more sensitive to thermal and photooxidation than PE alone
(due to higher oxidizability of PP), see inset Fig. 10c, and the weatherability of

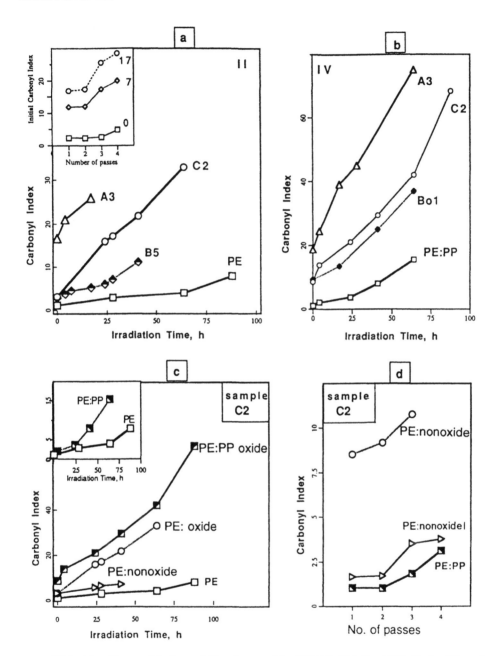

FIG. 10.   Photooxidative stability (exposed in SEPAP 12-24) of PE (Series II) and PE:PP (Series IV) blends containing 10% photooxidized degradables processed for one pass (a–c; in c the effect of PE containing nonoxidized C2 is also shown). Effect of multiple processing on initial carbonyl index of PE:PP blends containing oxidized and nonoxidized C2 is shown in d.

these blends (containing photooxidized degradables) is consequently more severely affected compared to PE containing the same photooxidized degradable samples, see Fig. 10c. These results show that while thermal reprocessing (recycling) of plastics waste containing small amounts of degradables can still be performed, especially if the process temperature is controlled, recyclates containing photooxidized degradables suffer more severely and will, in general, benefit from the use of additional photoantioxidants (and possibly thermal antioxidant) for improved outdoor performance as was demonstrated in the case of blends containing the degradable sample B5 which contains a very small concentration of a photoantioxidant.

CONCLUSIONS

1. Photooxidation of all degradables examined leads to the build-up of high concentrations of carbonyl compounds concomitant with a reduction in molecular weight. The rate of photooxidation depends on the sensitizing mechanism: starch-filled PE (containing no FeSt) showing the lowest rate and starch-FeSt-containing polymer showing much higher rates comparable to FeDRC-containing polymer. All the degradables showed faster photooxidation than a control LDPE (with no additives) and without an induction period except for the case of Fe/NiDRC (Scott-Gilead polymers) which gave an IP but without adversely affecting the subsequent rapid photooxidation which leads to embrittlement. In all cases the initial abiotic degradation is essential for the subsequent biodegradation process.

2. The rate of reduction in molecular weight of all degradables during the initial stages of UV exposure was rapid but slowed down at later stages; the ECO polymer, A3, showed the fastest initial rate of reduction in MW compared to the slower and more progressive reduction in the case of the iron-containing polymers (B1, B5, C2).

3. Photooxidation imparts hydrophilicity to the initially hydrophobic LDPE-based degradables. The low molecular weight photofragmented polymers were shown to contain high concentrations of carboxylic acids and, to a lesser extent, esters, ketones, and alcohols. When photofragmented to the same level of reduction in MW, the ECO polymer (A3) gave only about two-thirds of the total carbonyls formed in an FeDRC-containing polymer (B01) but showed a ninefold increase in vinyl absorption. This leads to the conclusion that photooxidation of A3 is dominated by a Norrish II photolysis of ketones and that its degradation is dependent on the initial amount of carbonyl available, while the Norrish I process is more important in the case of B01 with the formation of carboxylic acids as the major final degradation product. The lower total level of carbonyls in the photofragmented ECO polymer may account for the slower biodegradation observed compared to FeDRC-containing polymer.

4. Homogeneous and heterogeneous polymer blends can be recycled in the presence of a small concentration of nonoxidized degradables, and the adverse changes in polymer properties can be kept to a minimum by controlling the recycling conditions, e.g., temperature. The Fe/NiDRC-containing degradable polymers do not show adverse effects on recyclate properties both during processing and weathering due to the excellent antioxidancy of the dithiocarbamates. However, if plastics

waste becomes contaminated with photooxidized degradables, then both the processing and weathering stability of the recyclates will be severely affected, and antioxidants should be added during the recycling process to ensure long term performance of the blends in subsequent lives.

ACKNOWLEDGMENTS

The authors are grateful for the financial support received for this work from the European Commission via BRITE-EURAM project BE-3120-89. BP Chemicals, UK, and Exxon chemicals (UK) are acknowledged for supplying free samples of LDPE powder and EPDM, respectively.

REFERENCES

[1]   S. Al-Malaika and G. Scott, in *Degradation and Stabilisation of Polyolefins* (N. S. Allen, Ed.), Applied Science Publishers, London, 1983, Chap. 6.
[2]   S. Al-Malaika, in *Atmospheric Oxidation and Antioxidants, Vol. I* (G. Scott, Ed.), Elsevier Science Publishers, Amsterdam, 1993, p. 77.
[3]   *The Green Report*, compiled by a working party of the National Association of Attorneys General of the USA, 1990.
[4]   BRITE-EURAM EEC Programme, Project No. 3120, 1990, unpublished work.
[5]   G. Scott, in *Proceeding of the Third International Workshop on Biodegradation*, Osaka, Japan, 1993.
[6]   M. M. Brubaker, US Patent 2,495,286 (1950).
[7]   J. E. Guillet, in *Polymers and Ecological Problems* (J. Guillet Ed.), Plenum Press, New York, 1973, p. 1.
[8]   D. C. Mellor, A. B. Moir, and G. Scott, *Eur. Polym. J., 9*, 219 (1973).
[8a] M. U. Amin and G. Scott, *Ibid., 10*, 1019 (1974).
[9]   G. Scott and D. Gilead, British Patent 1,586,344 (1978).
[10] G. J. L. Griffin, British Patent Int. Publ. W088/09354 (1988) (Starch + Fe).
[11] G. J. L. Griffin, *J. Polym. Sci., Polym. Symp., 57*, 281 (1976).
[12] J. Lacoste and D. Carlsson, *J. Polym. Sci., A*, p. 493 (1992).
[13] A. Tidjani and R. Arnaud, *Polym. Degrad. Stab.*, In Press.
[14] S. Al-Malaika and G. Scott, in Ref. 1, Chap. 7.
[15] S. Al-Malaika, A. Marogi, and G. Scott, *J. Appl. Polym. Sci., 31*, 685 (1986).
[16] S. Al-Malaika, K. B. Chakraborty, and G. Scott, in *Developments in Polymer Stabilisation, Vol. 6* (G. Scott, Ed.), Elsevier Applied Science, 1983, Chap. 3.
[17] G. H. Hartley and J. Guillet, *Macromolecules, 1*, 169 (1968); F. J. Golemba and J. E. Guillet, *Ibid., 5*, 63 (1972).
[18] S. Al-Malaika, A. Marogi, and G. Scott, *Polm. Degrad. Stab., 18*, 89 (1987).

[19] S. Al-Malaika, A. Marogi, and G. Scott, *J. Appl. Polym. Sci., 33*, 1455 (1987).

[20] G. Scott, in *Biodegradable Polymers and Plastics* (M. Vert, J. Feijen, A. Albertsson, and E. Chielini, Eds.), Royal Soc. Chem., London, 1992, p. 291.

[21] M. Kramer, in *Proceedings of the Compoloy-Europe '92*, Shotland Business Research, Brussels, 1992, p. 275.

[22] M. U. Amin, G. Scott, and M. K. Tillekeratne, *Eur. Polym. J., 11*, 85 (1975).

CAN POLYETHYLENE BE A PHOTO(BIO)DEGRADABLE SYNTHETIC POLYMER?

JACQUES LEMAIRE, RENÉ ARNAUD, and PHILIPPE DABIN

Laboratoire de Photochimie Moléculaire et Macromoléculaire
URA CNRS 433
Ensemble Universitaire des Cézeaux
F-63177 Aubiere Cedex, France

GERALD SCOTT, SAHAR AL-MALAIKA, and SUKHINDER CHOHAN

Department of Chemical Engineering and Applied Chemistry
Aston University
Birmingham, UK

ANNIE FAUVE and ABDERRAZEK MAAROUFI

Laboratoire de Chimie Organique et Biologique
URA CNRS 485
Ensemble Universitaire des Cézeaux
F-63177 Aubiere Cedex, France

ABSTRACT

A polymeric system based on LDPE would be qualified as a "photo(bio)degradable synthetic polymer" for use as films or thin systems in plasticulture and later in packaging where severe specific criteria should be respected. The evolution of such a system in environmental conditions should present three phases. In Phase I, corresponding to storage and use, in the presence of physicochemical and biological aggression, chemical evolution should be very limited and resistance to any microorganism should be observed. In Phase II a rapid abiotic degradation should occur until the complete destruction of physical (mechanical) properties and spontaneous fragmentation of the thin systems into more and more di-

vided parts. Phase III corresponds to bioassimilation of heavily transformed (oxidized) solid particles. Phase I should be predicted and controlled on the basis of artificial photoaging or thermoaging experiments.
Depending on the desired lifetime of the system, nonaccelerated, accelerated, or ultra-accelerated photoaging techniques could be used. The earliest fragmentation, which should be observed in Phase II, should be
predicted within the same experiment. The prediction of the long-term
fate of the polymeric materials should be based not only on the variations of physical properties but on a full analysis of the chemical evolution, i.e., determination of the major final transformed groups of the
macromolecular chains (and especially the acidic end groups) and the
molar mass distribution. In a recent BRITE-EURAM European contract, we developed an experimental protocol for the control of Phase III
based on the use of pure cultures of strains from collections or selected
adapted wild strains (from industrial polyethylene site dumpings) which
had been examined. Abiotically oxidized LDPE was the only carbon
source in a starving mineral medium. Bioerodibility caused by the carboxylic acid formed throughout abiotic degradation has been observed.

INTRODUCTION

In response to public concern about the effects of plastics in the environment,
and in particular the damaging effects of sea litter on animals and birds, legislation
is being enacted or is pending in many countries to ban nondegradable packaging,
fishing nets, etc. In response to these concerns a variety of new polyolefin-based
packaging materials have recently appeared on the market which claim to be photodegradable, biodegradable, or both. A problem for the manufacturers and users of
plastics products, and ultimately for legislators, is the absence of objective criteria
to distinguish between the performance of competing products and processes.

In a recent European research contract (BRITE EURAM BREU 170) entitled
"An Investigation of the Parameters Involved in the Environmental Ultimate Degradation of Plastics," Université Blaise Pascal and Aston University collaborated on a
program whose main objectives were to provide:

1. Guidance for the manufacturers and users of degradable plastics which will
    allow them to distinguish between the claims of different producers
2. The basic principles underlying the developments of industrial standards for
    quality and assurance and for quality control

The research was carried out on most types of "photodegradable" polyethylene
systems commercially available and on fully controlled on-purpose-made films. As
later reported, the experimental results collected in the 3-year cooperation appear as
a first approach to answer the general question: "Can polyethylene be a photo(bio)-
degradable synthetic polymer?"

DEFINITION OF A PHOTO(BIO)DEGRADABLE SYNTHETIC POLYMER

A polymeric system based, for example, on LDPE or LLDPE would be qualified as a photo(bio)degradable synthetic polymer for use as films or thin systems in

plasticulture or in the packaging industry when the following severe specific criteria are respected. The evolution of such a system should present three phases.

Phase I corresponds to storage in usual conditions where the physicochemical strains could be limited to O_2, heat corresponding to room temperature, humidity, and microorganisms. Such storage conditions are usual in the plasticulture industry. However, if extension to the packaging industry is considered, the environmental strains during Phase I should include daylight (more or less filtered according to external or internal use). Any packaging systems should maintain their initial physical properties (including permeability) during use, even in the presence of daylight. Chemical evolution in Phase I, should be very limited as should variations in macroscopic physical properties and resistance to any microorganisms. Phase I should extend for several months in plasticulture and from several months to more than 1 year in packaging.

In *Phase II* a rapid abiotic degradation should proceed until the complete destruction of mechanical properties is attained, and there should be spontaneous fragmentation of thin systems into more and more divided parts. Such an abiotic degradation results from the concerted action of daylight, heat, and O_2. Water as humidity does not affect LDPE and LLDPE photothermal oxidation, and biodegradation is not required at this level.

Phase III corresponds to the bioassimilation of heavily transformed (oxidized) solid particles.

Phase I should be predicted and controlled on the basis of artificial photoaging or thermoaging experiments. Depending on the desired lifetime of the system, non-accelerated, accelerated, or ultra-accelerated photoaging techniques could be used. The earliest fragmentation, which should be observed in Phase II, should be predicted within the same experiment. The prediction of the long-term fate of the polymeric materials should be based not only on the variations of physical properties but on a full analysis of the chemical evolution, i.e., the determination of the major final transformed groups of the macromolecular chains (and especially the acidic end groups) and the molar mass distribution.

The control of Phase III, i.e., evaluation of the biodegradability of heavily oxidized LDPE fragments, requires definition of an uncontroversial protocol, which is far beyond the present status of the art. An experimental protocol based on the use of pure cultures of strains from collections or selected adapted wild strains (from industrial polyethylene site dumpings) will be presented in the next sections. Abiotically oxidized LDPE was the only carbon source in a starving mineral medium.

CONTROLLED ABIOTIC DEGRADATION OF LDPE SYSTEMS

It has been recognized during the past 20 years that for some applications, particularly in packaging and in some agricultural applications such as mulching films and binder twines, the polyolefins as normally formulated are too stable for their intended purposes. This has led to the development of commercial products with enhanced degradability. The following classes of "degradable polyolefins" (i.e., abiotically fragmentable) have been identified.

Photolytic polymers (type A) contain a carbonyl group which is either built into the polymer chain (e.g., ethylene–carbon monoxide polymers) or they contain a copolymerized vinyl ketone in the side chain (the Guillet process) [1]. Both modifications undergo photolysis by the Norrish II (and to a lesser extent the Norrish I) process with chain scission.

Polymers with enhanced photooxidability (type B). There are two subdivisions of this class. 1) Polymers which contain an organo-soluble metal ion (generally in the form of a carboxylate), which acts as a thermal- and photoprooxidant for the polymer. 2) Polymers which contain a sulfur complexed metal ion [e.g., Fe(III) dithiocarbamate] which is photolyzed during outdoor exposure, liberating the free metal ion which then assumes its normal prooxidant function (Scott–Gilead process) [2].

Starch-filled polyethylene (type C). Again, there are two modifications of this system (the Griffin process) [3]. 1) This polymer contains only starch, normally at a 6–8% concentration. 2) This polymer contains, in addition to starch, an iron carboxylate (generally stearate), which, as in B(1), catalyzes the photooxidation of the polymer, thus releasing the starch filler.

CONTROL OF PHASE I IN STORAGE CONDITIONS

In the absence of light, LDPE and LLDPE are only involved in room temperature thermooxidation. As reported later, among the 70 strains examined in the bioassimilate experiments, not one was able to grow on unoxidized LDPE. Considering the thermal stability of the main intermediate and first oxidation products (associated ROOH, isolated ROOH, ketonic groups, alcoholic groups, acidic groups), the thermooxidation mechanism observed at room temperature should be unchanged up to 90–100°C, and the apparent activation energy should be constant in the 20–100°C range.

Thermooxidability of the various systems was examined in an oven under constant air flow or without constant air flow between 90 and 110°C. With the exception of starch-filled LDPE films, all other samples exhibit an induction period (400–600 hours at 80°C, 200–300 hours at 100°C). From the determined activation energy, it is possible to predict that significant oxidative conversion should not be observed before 100–200 days for the various systems at 20°C.

The limitation of chemical evolution or induction period which should be observed in Phase I when such a system is used in the presence of light is controlled in the same experiment as Phase II.

CONTROL OF PHASE II

Various experimental conditions have been used.

A nonaccelerated photoaging cabinet designed in Aston University, using 7 sunlamps and 21 actinic blue lamps arranged in a symmetrical sequence. The maximum in relative intensity was found to occur within the range 280–370 nm, and the average relative radiation output in the cabinet is estimated to be in the 4.5 $W \cdot h \cdot m^{-2}$ range.

An accelerated photoaging SEPAP 12-24 unit was designed in Laboratoire de Photochimie in Clermont-Ferrand. This device is fitted with four 400-W medium-pressure mercury lamps located at the four corners of a square chamber. The samples were irradiated on a rotating support located at the center of the chamber. The radiations were filtered at $\lambda > 300$ nm, and the temperature of the samples was 60°C.

An ultra-accelerated photoaging SEPAP 50-24 unit was designed in Laboratoire de Photochimie in Clermont-Ferrand. This device is similar to the SEPAP 12-24 unit, but it includes 8 lamps and operates at 70°C.

The "mechanism" of LDPE photooxidation is fairly well established [4, 5]. The photochemical evolution of LDPE can be described through variations in the concentration of acidic end groups, considered as critical photoproducts and determined by FT-IR spectrophotometry as compounds absorbing at 1710–1715 cm^{-1}. The samples were exposed to UV radiation until complete degradation was reached, i.e., the fragmentation under minor strain usually observed when the absorbance at 1715 cm^{-1} is close to 0.3 for 30 μm films.

As an example, the variations of the absorbance at 1715 cm^{-1} for A_2, B_5, and C_2 films exposed in a SEPAP 12-24 unit are presented in Fig. 1 and compared to the corresponding variations of an untreated PE film.

GPC measurements have been carried out with a Waters unit, model 150C ALC/GPC. 1,2,3-Trichlorobenzene containing 250 ppm of Santonox R, a phenolic antioxidant, has been used as the solvent at 135 and 160°C. Prior to any GPC run, the complete dissolution of the unoxidized or photooxidized LDPE samples was obtained when the solvent was heated at 160°C for 4 hours. The solution was thoroughly filtered. The GPC runs were carried out at 135°C, the Waters unit being equipped with four μ-Styragel columns (pore sizes 10^3, 10^4, and 10^5Å ; particle size 10 μm; length 30 cm;

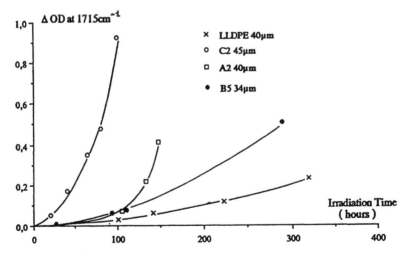

FIG. 1.   Variations of absorbance at 1715 cm^{-1} vs exposure time for A_2, B_5, and C_2 films and untreated LDPE film samples (SEPAP 12-24 unit, 60°C).

diameter 7.8 mm). Detection was based on refractometry. For calibration, TSK monodisperse polystyrene standards from Varian were used after dissolution in trichlorobenzene at 145°C in less than 1 hour and requiring no filtering.

A direct comparison of the rates of photooxidation of commercial degradable polyethylenes showed a rapid rate of carbonyl formation and a molar mass reduction for all the polymers which were claimed to be photodegradable. Three different accelerated weathering devices (S/B cabinet, SEPAP 12-24, SEPAP 50-24) were used, and a good correlation was found between them and outdoor exposures.

From the results reported in Table 1, it appears that the changes of the absorbance at 1715 cm^{-1} can be related to the variations of the average molar mass ($\overline{M}_n$, $\overline{M}_w$). An oxygen starvation effect was not observed in thin films (12–45 μm) which behaved like a homogeneous photoreactor, whatever the acceleration conditions were.

It is quite clear that UV exposure under all conditions led to a considerable reduction in the molar mass ($\overline{M}_n$, $\overline{M}_w$). Note the following important points:

Values of $\overline{M}_w$ at embrittlement for the photooxidizing polymers were generally in the 10,000–20,000 region ($\overline{M}_n$ = 5,000–10,000); these samples were subsequently studied in biodegradation.

As shown, for example, on the C$_2$ films, in ultra-accelerated, medium accelerated, and nonaccelerated conditions, a very similar relationship between the reduction of molar mass and the build up of the acidic groups was observed (see Fig. 2). This means that even in the presence of the highest concentrations of radicals, crosslinking never competes in chain scissions throughout the photooxidation of C$_2$ films.

Molar mass reduction was particularly rapid during the initial stages of UV exposure in the case of the ethylene–carbon monoxide copolymer (A$_2$), but the degradation (as measured by this parameter) apparently ceased at a limiting $\overline{M}_w$ of about 50,000, whereas all the transition metal systems continued to much lower values of $\overline{M}_w$. This behavior of the photolytic polymers has been reported previously [6], and it has been suggested that crosslinking through vinyl competes with chain scission at a relatively early stage in photolysis [7].

CONTROL OF PHASE III: BIODEGRADATION STUDIES

Assuming that Phase II abiotic oxidation of polyethylene may favor the biotic attack of the matrix by creating assimilable long-chain carboxylic acids, as already reported by several authors (cf., for example, Refs. 8 and 9), a comparative study of the bioassimilation of nonoxidized and oxidized plastic films was carried out. After selection of the microbial species from collections or from polyethylene-rich areas, various types of commercial photo(bio)degradation polyethylene films were incubated under controlled starving conditions. Besides molecular weight modification analysis by GPC and FT-IR of the matrix, microbial action was especially monitored by scanning electron microscopy analysis of the surface.

TABLE 1.

Films references	Exposure conditions	e, μm [b]	Exposure time	$\Delta_{absorbance}$ at 1715 cm^{-1}	M_n [a] ($\pm 5\%$)	M_w [a] ($\pm 5\%$)
A$_2$	SEPAP 12-24 (60°C)	12	–	–	72,500	291,400
			85 hours	≈ 0	17,300	131,900
			105	0.020	9,600	48,800
			133	0.065	7,400	40,400
			148	0.124	6,800	36,600
B$_5$	SEPAP 12-24 (60°C)	34	–		33,200	229,500
			26 hours	0.008	37,600	214,500
			93	0.063	17,400	143,700
			110	0.077	13,700	113,000
			289	0.508	5,100	29,300
	S/B Cabinet	35		2.132	1,800	4,500
	Weathering	34	–	–	33,200	229,500
	(45° south)		23 days	0.142	53,200	239,300
			40	0.173	26,800	208,800
			86	0.305	14,300	105,100
			124	0.428	10,700	63,600
C$_2$	SEPAP 12–24 (60°C)	45	–	–	55,000	271,000
			20 hours	0.05	31,690	116,300
			40	0.17	12,100	78,600
			65	0.35	13,900	52,100
			80	0.48	11,500	43,000
			100	0.924	3,500	13,800
			–	0.959	2,700	11,000
	SEPAP 50-24 (80°C)	45	–	–	55,000	271,000
			11 hours	0.142	16,090	58,100
			33	0.173	13,500	51,500
			–	0.305	7,500	34,200
			41	0.428	6,600	26,300
			–	0.618	4,900	16,700
	S/B Cabinet	45		2.075	2,600	8,500
	Weathering	45	–	–	55,000	271,000
	(45° south)		23 days	0.113	20,000	125,900
			40	0.249	14,000	76,700
			86	0.630	6,400	27,100

[a]Molecular weight in equivalent weight of polystyrene.
[b]e is the film thickness.

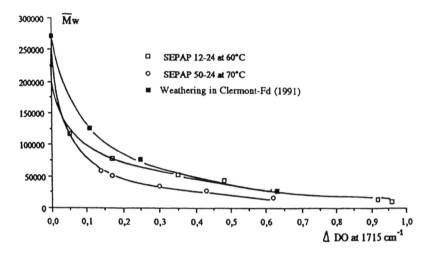

FIG. 2. Variations of M_w vs variations of the absorbance at 1715 cm^{-1} in various photoaging conditions.

Selection of Strains

Biodegradation of long-chain molecules is well documented, and 32 different species were tested first in the selection procedure. Plastic-degrading strains, rubber-degrading strains, and hydrocarbon-degrading strains used for waste degradation were tested. They were mainly bacteria and fungi with simple nutritional requirements, nonpathogenous, and easily available from collections. Evolution of the microbial population according to chemical modifications is generally observed in the environment. Adapted species may be found in polyethylene-rich areas, and four different locations were investigated. Considering the enormous range of microorganisms in soil, a selective technique, the enrichment culture procedure, was used. In the procedure, oxidized polyethylene films were the growth-limiting and sole carbon source, and only organisms with the necessary polyethylene-degradative ability could grow and outgrow the large number of strains added at the start. Thirty-eight pure cultures were isolated and purified. They were mainly fungi and bacteria, and they were fully characterized after selection.

Pure cultures of microorganisms were incubated in a classical mineral medium in which the only carbon source was the plastic film. Photodegraded copolymer A_2, photosensitive B_5, and iron stearate-starch containing C_2 were used as the carbon sources in a starving medium in which all nutrients were minerals. Samples were sterilized and inoculated with each of the 70 selected strains. After 6 months of incubation in weathering ovens, the microbial growth was observed by visual rating and comparison with various controls (nonoxidized films, abiotic and viability controls). Only strains which displayed a visual growth ranking of at least 2 on a scale of 4 for the three types of films were selected. After analysis of the chemical modifications as described further, the following strains were selected:

Cladosporium cladosporioides (no. 9), a fungus from the American Type
   Culture Collection (20251)

Norcardia asteroides (no. 13), a bacterium isolated and characterized earlier
   in our laboratory (911)
Rhodococcus rhodochrous (no. 28), a bacterium from the American Type
   Culture Collection (29672)

Adapted species (two strains of the same fungus *Aspergillus* isolated at the
surface of plastic films and five isolated from a factory dumping site) were also
studied. However, their biodegradative abilities will not be fully presented here.

Bioassimilation of Commercial Degradable Films

Comparative bioassimilation studies of nondegraded, photodegraded, and
thermodegraded films were carried out. Emphasis was given to the modifications at
the surface of the film as shown by scanning electron microscopy observations.
Changes in polyethylene chemical structures were monitored through the variations
of carbonyl, carboxyl, and ester end-groups detected by FT-IR, through molecular
weight distribution analysis by GPC, and through the decrease of thickness, also
measured by FT-IR at 1375 cm^{-1}

Model films in each series were also A_2, B_5, and C_2. Because the inherent
heterogeneity of the polymers can lead to false conclusions, we used a procedure in
which the blank (zero incubation), control (abiotic incubation), and assay (biotic
incubation) test materials were cut from the same sample. Films were photodegraded in an SEPAP unit until the first stage of embrittlement was observed. Ther-
modegradation was performed at 90°C in an oven without a constant air flow. No
embrittlement appeared after weeks of thermal treatment, whereas oxidation had
occurred. The solid incubation medium was all-mineral. Assays and controls were
incubated 6 months with all selected pure strains and harvested every 2 months.
Because the thermal treatment was too long for A_2 and B_5 films, comparison by
oxidative treatment for the three diffferent matrixes was made with only one strain,
Nocardia; only C_2 was incubated with the three strains after both photo- and ther-
moxidation. Biodegradation was evaluated in terms of oxidation treatment and
polymer content, microbial species, and incubation time.

As-received nondegraded photodegradable LDPE films (M_w = 300,000) were
not attacked by the microbial species tested. By looking at SEMs of incubated
samples it appeared that abiotic degradation due to the incubation conditions was
not significant compared to the modifications due to microbial cells observed at
the surface of degraded films. A biotic degradation of the surface of photoxidized
and/or thermooxidized samples was evidenced, especially for B_5 and C_2 films
(cf. Table 2).

GPC analysis showed a slight decrease of average M_w after biotic incubation,
but this reduction was not significant enough to be used as a quantitative predictive
technique in biodegradability studies. Data obtained by FT-IR spectra analysis at
1375 cm^{-1} showed a decrease of thickness as determined by a decrease of absorbance, and this was used to quantify the biodegradability of the films.

SEM observations and FT-IR measurements of the surface erosion showed
that:

No strain was found to assimilate nonoxidized material.
Only one strain, *Nocardia*, was found to modify copolymer A_2 and then only

TABLE 2.   Optical Density and Mass Distribution Values of A_2, B_5, and C_2 Films vs Oxidation Treatment after 6 Months of Incubation with *Nocardia*, Blanks, Abiotic Controls, and Assays

Film	Abiotic treatment	Blank O.D. 1715 cm^{-1}	Blank M_w ($\times 10^3$)	Abiotic control O.D. 1715 cm^{-1}	Abiotic control M_w ($\times 10^3$)	*Nocardia* no. 13 O.D. 1715 cm^{-1}	*Nocardia* no. 13 M_w ($\times 10^3$)
A_2	Nontreated	0.05	280	0.05	310	0.04	320
	Photodegraded	0.12	17	0.10	16	0.11	12
	Thermodegraded	0.45	21	0.40	19	0.20	17
B_5	Nontreated	0	250	0	250	0	250
	Photodegraded	0.35	40	0.30	32	0.25	19
	Thermodegraded	1.05	16	0.90	16	0.75	15
C_2	Nontreated	0.10	200	0.11	280	0.11	280
	Photodegraded	1.15	16	1.08	15	0.70	12
	Thermodegraded	1.90	ND	1.95	ND	1.55	ND

after thermal treatment. In general, it was observed with all strains, even the adapted species, that this degraded copolymer A_2 was not a good substrate for biodegradation at the embrittlement stage. This may be because, in this series, photolytic fragmentation according to a Norrish II type reaction did not afford, at this early stage, the long-chain carboxylic acids assumed to favor microbial assimilation.
Strains of particular genera, especially *nocardioforms* and *aspergilli*, from collections or adapted sites, strongly alter B_5 and C_2.

Photosensitive film B_5 was a good substrate for microbial growth after photo- and thermooxidation, depending on the species used. Iron stearate-starch-containing photo(bio)degradable C_2 film was an even better substrate for microbial growth after oxidation. All the strains tested could assimilate starch, thus increasing the rate of growth. However, this biodegradable additive must be released from the polymer matrix by abiotic treatment prior to incubation.

The kinetics of bioassimilation were monitored for each *phototreated* model film with the selected strains. Film samples were analyzed as already reported, and the results varied according to the type of film:

Untreated films showed no modifications within 6 months of incubation.
Phototreated A_2 presented no modifications after 6 months of incubation with all strains. After 4 months, thermotreated A_2 was eroded (-4%).
Bioassimilation of photooxidized B_5 was observed after 4 months of incubation with *Rhodococcus* whereas 6 months were required for the measurement of a decrease of thickness of the film with the other efficient strains (Table 3).
Photooxidized starch-containing film C_2 was rapidly covered with a biomass,

TABLE 3. Bioassimilation of Photooxidized B_5 and C_2 Films vs Time. Decrease of Thickness vs Controls and Blanks

Microorganisms	Films	2 months, %	4 months, %	6 months, %
Cladosporium no. 9	A_2	0	0	0
	B_5	0	0	−6
	C_2	0	−13	−27
Nocardia no. 13	A_2	0	0	0
	B_5	0	0	0
	C_2	0	−33	−27
Rhodococcus no. 28	A_2	0	0	0
	B_5	0	−3	−15
	C_2	−4	−5	−11

and a decrease of thickness was observed after 2 months of incubation with *Rhodococcus* and 4 months with *Cladosporium* and *Nocardia*.

CONCLUSIONS

LDPE-based photodegradable systems should be considered to be photo(bio)-degradable. They undergo abiotic degradation which can be programmed and predicted from artificial aging experiments. After fragmentation, the heavily oxidized particles have erodibility in the presence of several selected strains in starving mineral media where the degraded polymer is the only carbon source. This result contrasts with those obtained in nonstarving media.

REFERENCES

[1] J. E. Guillet, US Patents 3,753,952, 3,811,931, 3,860,538, and 3,878,169.
[2] G. Scott, in *Polymers and Ecological Problems* (J. Guillet, Ed.), Plenum, 1973, p. 27.
[3] G. J. L. Griffin, in *Proceedings of the Symposium on Degradable Plastics*, SPI, Washington, D.C., 1987, p. 49.
[4] M. H. Amin, G. Scott and L. M. K. Tillikeraine, *Eur. Polym. J., 11*, 85 (1975).
[5] R. Arnaud, J. Y. Moisan, and J. Lemaire, *Macromolecules, 17*, 332 (1984).
[6] R. J. Statz and M. C. Dorris, in *Proceedings of the Symposium on Degradable Plastics*, SPI, Washington, D.C., 1987, p. 51.
[7] G. Scott, *Polym. Degrad. Stab., 29*, 135 (1990).
[8] A. C. Albertsson, S. O. Andersson, and S. Karlsson, *Ibid., 18*, 73 (1987).
[9] A. C. Albertsson, *J. Macromol. Sci. — Pure Appl. Chem., A30*, 757 (1993).

BIODEGRADABLE POLYURETHANES FROM PLANT COMPONENTS

HYOE HATAKEYAMA, SHIGEO HIROSE, and TATSUKO HATAKEYAMA

National Institute of Materials and Chemical Research
1-1, Higashi, Tsukuba, Ibaraki 305, Japan

KUNIO NAKAMURA

Otsuma Women's University
Chiyoda-ku, Tokyo 102, Japan

KEN KOBASHIGAWA

Tropical Technology Center Ltd.
Gushikawa, Okinawa 904-22, Japan

NORIYUKI MOROHOSHI

Tokyo University of Agriculture and Technology
Fuchu, Tokyo 183, Japan

ABSTRACT

Polyurethane (PU) sheets and foams having plant components in their network were prepared by using the following procedure. Polyethylene glycol (PEG) was mixed with one of the following; molasses, lignin, woodmeal, or coffee grounds. The mixture obtained was reacted with diphenylmethane diisocyanate (MDI) at room temperature, and precured PUs were prepared. The precured PUs were heat-pressed and PU sheets were obtained. In order to make PU foam, the above mixture was reacted with MDI after the addition of plasticizer, surfactant (silicone oil), catalyst (di-n-butyltin dilaurate), and droplets of water under vigorous stirring. The glass transition temperature, tensile and compression strengths, and Young's modulus of the PU sheets and foams increased with an increasing amount of plant components. This suggests that saccharide and lignin residues act as hard segments in PUs. It was found that the PUs obtained were biodegradable in soil. The rate of biodegradation of the PUs derived from molasses and coffee grounds was between that of cryptomeria (*Cryptomeria japonica*) and beech (*Fagus sieboldi*).

INTRODUCTION

Natural polymers are materials which are highly adaptable to their circumstances. They are also materials having appropriate reactivity because of functional groups such as hydroxyl groups. Accordingly, natural polymers having more than two hydroxyl groups per molecule can be used as polyolys for polyurethane (PU) synthesis.

Synthetic polymers, which contain natural polymer or their main components such as carbohydrate and lignin residue, are considered to be biodegradable. Therefore, extensive studies have been carried out to synthesize polyurethanes (PUs) which were derived from plant components [1–5]. In these studies, PUs were prepared using polyols such as polyethylene glycol (PEG) and polypropylene glycol (PPG).

The present paper reports our recent studies on the thermal and mechanical properties, and also the biodegradability of PUs prepared from molasses and lignocellulose, such as lignin, woodmeal, and coffee grounds, by polymerization with PEG, PPG, and diphenylmethane diisocyanate (MDI).

EXPERIMENTAL

Materials

Molasses (ML) were supplied by Tropical Technology Center Ltd. Lignin (Kraft lignin, KL) was obtained from Holmen Pulping Company. KL was purified by the precipitation method using dilute sulfuric acid. Hardwood solvolysis lignin (SL) was obtained as a by-product in organosolve-pulping of Japanese beech (*Fagus crenata*). The SL was provided by the Japan Pulp and Paper Research Institute Co. Woodmeal from pine was supplied by Miki Sangyo Co. Ltd. The particle size was 60–90 mesh. Coffee grounds were provided by Ueshima Coffee Co. Ltd. PEG; PPG and MDI were commercially obtained. Wood blocks of cryptomeria (*Cryptomeria japonica*) and beech (*Fagus sieboldi*) were obtained from the Experimental Forest Station which belongs to Tokyo University of Agriculture and Technology.

Preparation of PUs

In order to obtain PUs, it was necessary to dissolve or suspend molasses, lignin, woodmeal, and coffee grounds in polyols such as PEG and PPG. The polyol solutions (in the case of molasses and lignin) or suspensions (in the case of woodmeal and coffee grounds) obtained were mixed with MDI at room temperature, and precured PUs were prepared. Each of the precured PUs was heat-pressed and a PU sheet was obtained. In order to prepare PU foams, the above polyol solution was mixed with plasticizer, surfactant (silicone oil), and catalyst (di-*n*-butyltin dilaurate), and then MDI was added. The mixture was vigorously stirred with droplets of water which were added as a foaming agent, and a PU foam was obtained.

Measurements

Thermal properties of PU samples were measured using a differential scanning calorimeter, Seiko DSC 220, and a thermogravimeter, Seiko TG 220, with a workstation 5700. DSC and TG curves were obtained at the scanning rate of 10°C/min.

Mechanical properties of PU samples were measured using a tensile test machine, Shimadzu Autograph 500. The strain rate was 5 mm/min for the tensile test and 2 mm/min for the compression test.

Biodegradation

PU foams derived from molasses and coffee grounds were cut into specimens of 5 cm (width) × 5 cm (length) × 1 cm (thickness). The specimens were pinned on the surface of the ground at the hilly farm that belongs to the Faculty of Agriculture of Tokyo University of Agriculture and Technology, and covered lightly with soil and a net, and they were kept in this state for predetermined periods. Then a certain number of the specimens were taken from the farm, washed, and dried. The average weight loss of five specimens of each kind of sample was calculated according to the following equation: Weight loss (%) = $(W_s - W_d)/W_s$, (g/g) × 100, where W_s (g) is the sample weight and W_d (g) is the sample weight after the sample was kept in the soil for a certain time.

RESULTS AND DISCUSSION

Chemical Structure of Prepared PUs

The core structure of prepared PUs consists of saccharide and lignin linked by urethane bonding. Accordingly, it can be assumed that the PUs obtained are essentially copolymers having three-dimensional networks of urethane bonding which have saccharide and lignin components combined with polyol (PEG or PPG). Schematic chemical structures of PUs derived from saccharide and lignin are shown in Fig. 1.

Thermal Properties of PUs

Figure 2 shows DSC curves of PU sheets with various molasses contents in the PEG–ML system. A marked change in baseline due to glass transition was observed in each DSC curve. Glass transition temperatures (T_gs) were determined by a method reported previously [6]. T_g increased with increasing content of molasses in polyol, since pyranose and franose rings from saccharides in molasses act as hard segments in PU networks.

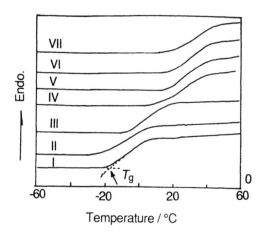

FIG. 1.   Schematic chemical structures of PUs from saccharides and lignin.

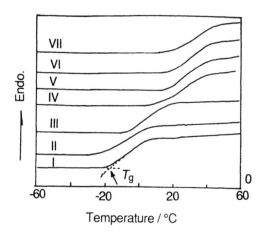

FIG. 2.   DSC curves of PU sheets with various molasses contents in PEG–molasses systems. Molasses contents (%): I, 0; II, 5; III, 10; IV, 15; V, 20; VI, 25; VII, 30. Heating rate = 10°C/min.

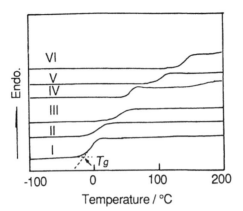

FIG. 3. DSC curves of PU sheets with various lignin (SL) contents in PEG–SL systems. SL contents (%): I, 0; II, 10; III, 20; IV, 30; V, 40; VI, 50. Heating rate = 10°C/min.

Figure 3 shows DSC curves of PU sheets with various lignin (SL) contents in the PEG–SL system. T_g increases with increasing SL content. It is known that lignin is a crosslinked and highly branched polymer which has rigid phenyl propane structures as repeating units. Lignin also has more than two hydroxyl groups in a molecule. Accordingly, it is considered that lignin reduces the mobility of the main chain of PU molecules.

A similar relationship was observed in the change of T_gs of PUs with various contents of KL, woodmeal, and coffee grounds. In all cases the increase of plant components in PUs resulted in the increase of T_g.

Mechanical Properties of PU Sheets

Figure 4 shows changes of stress (σ, MPa) and elongation (ϵ, %) at the breaking points of PU sheets with various molasses contents. As shown in the figure, the σ value increases with increasing molasses content in polyol. The value

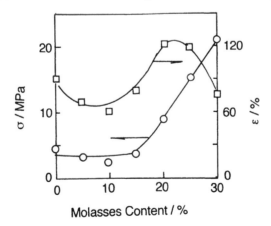

FIG. 4. Relationships between ultimate strength (σ) and ultimate strain (ϵ) of PU sheets and molasses contents in polyol systems.

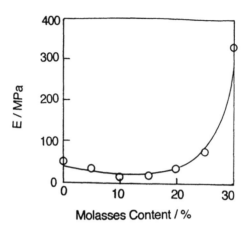

FIG. 5.   Relationship between Young's modulus (E) and molasses content in polyol systems.

shows a prominent increase when the molasses content exceeds 20% since the T_g value of the PU sheet becomes higher than room temperature (Fig. 2) and the PU molecules are in the glassy state. The ϵ value of the PU sheet derived from molasses shows a maximum at around the molasses content of 20%, since the molecular state of the PU sheet changes from the rubbery state to the glassy state at room temperature when the molasses content exceeds 20%.

As shown in Fig. 5, Young's modulus (E, MPa) of a PU sheet derived from molasses increases with increasing molasses content, since pyranose and franose rings from the saccharides in molasses act as hard segments in PU networks.

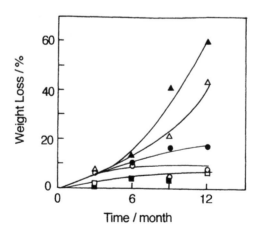

FIG. 6.   Relationship between weight loss and degradation time. ($\triangle$, $\blacktriangle$) Buna; ($\bigcirc$, $\bullet$) PU derived from molasses (33%); ($\square$, $\blacksquare$) Sugi; ($\triangle$, $\bigcirc$, $\square$) degraded at a hilly open farm; ($\blacktriangle$, $\bullet$, $\blacksquare$) degraded at a farm in a forest.

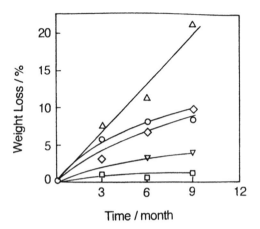

FIG. 7.   Relationship between weight loss and degradation time at a hilly open farm.
(△) Buna; (○) PU derived from coffee grounds (20%); (◇) PU (30%); (▽) PU (40%);
(□) Sugi.

Biodegradation of PU Foams

Figure 6 shows the weight loss (%) of PU foams derived from molasses which
were kept in soil for a certain period up to 12 months according to our experimental
plans. Figure 7 shows weight loss (%) of PU foams derived from coffee grounds
which were kept in soil for a certain period up to 9 months. In the above figures,
SUGI represents the weight loss of cryptomeria samples, BUNA that of beech
sample, MO 33% and MO 50%, those of PU foams prepared from polyols containing 33% and 50% molasses. A group of the specimen was biodegraded at a hilly
open farm, and the other group of specimens was biodegraded at a farm in a forest,
both of which belong to the Tokyo University of Agriculture and Technology. As
seen from the above figures, weight losses of PU foams derived from molasses are
lower than those of beech blocks but higher than those of cryptomeria. This suggests
that the PUs prepared in this study have good biodegradability, comparable with
natural wood.

CONCLUSION

The following conclusions are obtained from the above experimental results.

1. PU films and foams having a variety of thermal and mechanical characteristics
    were derived from molasses, lignin, woodmeal, and coffee grounds.
2. The glass transition temperature and the breaking strength of PU increased with
    increasing amounts of plant components in PU, since saccharide and lignin
    residues act as hard segments in the PU molecular networks.
3. The above PUs were biodegraded by microorganisms in soil, showing biode-
    gradability between that of beech and cryptomeria.

REFERENCES

[1] S. Hirose, S. Yano, T. Hatakeyama, and H. Hatakeyama, in *Lignin, Proper-
     ties and Materials* (ACS Symposium Series 397, W. G. Glasser and S. Sarka-
     nen, Eds.), American Chemical Society, Washington, D.C., 1989, p. 382.
[2] H. Yoshida, R. Morck, A. Reiman, K. P. Kringstad, and H. Hatakeyama, *J.
     Appl. Polym. Sci., 34*, 1187 (1987).
[3] K. Nakamura, R. Morck, A. Reiman, K. P. Kringstad, and H. Hatakeyama,
     Polym. Adv. Technol., 2, 41 (1991).
[4] K. Nakamura, T. Hatakeyama, and H. Hatakeyama, *Ibid., 3*, 151 (1992).
[5] H. Hatakeyama, K. Nakamura, S. Hirose, and T. Hatakeyama, in *Cellulosics;
     Chemical, Biochemical and Material Aspects* (J. F. Kennedy, P. A. Williams,
     and G. O. Phillips, Eds.), Ellis Horwood Ltd., Chichester, 1993, p. 525.
[6] S. Nakamura, M. Todoki, K. Nakamura, and H. Kanetsuna, *Thermochim.
     Acta, 163*, 136 (1988).

STARCH PROPERTIES, MODIFICATIONS, AND APPLICATIONS

J. JANE

Center for Crops Utilization Research and Department of
    Food Science and Human Nutrition
Iowa State University
Ames, Iowa 50011

ABSTRACT

Starch is the second largest biomass produced on earth. Starch consists mainly of amylose and amylopectin. The chemical structures of starch molecules, polymers of α-1-4 linked glucopyranose with α-1-6 branches, render a natural tendency for starch in aqueous systems to crystallize to double helices. Depending on botanical origin and genetic background, starch has different chemical structures, such as branch-chain lengths and phosphate derivatives, and different functional properties. This structural and functional diversity makes these starches suitable for different applications. High-amylose starch produces strong films and is suitable for making biodegradable plastics, and small-granule starch is an appropriate polyethylene film filler. Various physical, chemical, and enzymatic modifications will change and improve functional properties of starch and facilitate its utilization for different purposes. Pregelatinized and cold-water-soluble starches enable users to disperse starch without going through a heating process. Hydrophobic starch derivatives enhance the compatibility between starch and hydrophobic components. Dialdehyde starch facilitates crosslinkages between starch and other components carrying functional groups, such as amino, hydroxyl, and sulfurhydryl groups.

STARCH STRUCTURES AND PROPERTIES

Starch is the major form of carbohydrate storage in green plants and is considered the second largest biomass, next to cellulose, produced on earth. Starch is a polymer that consists of six-member-ring glucose units (glucopyranose). The molecular weight of starch varies from 10^4 to 10^7 daltons [1]. In contrast to cellulose, glucose units in starch are linked by α-1-4 bonds instead of the β-1-4 bonds found in cellulose. The α-1-4 linked starch displays a random-coil conformation, whereas the β-1-4 linked cellulose displays a ridged linear structure. The majority of starch molecules (e.g., up to 100% in waxy starches, 72% in normal maize starch, and 80% in potato starch) have highly branched structures, known as amylopectin [2]. Amylopectin has an α-1-4 linked backbone and ca. 5% α-1-6 linked branches. Branch chains in amylopectin are arranged in clusters [3, 4], and the branch-chain lengths vary with the species and genetic background of the starch. The branch chains are present in double helical, crystalline structures [3–5]. Amylose is a primarily linear molecule that has few branches [6]. In certain high-amylose varieties, the amylose content can be as high as 50 or 70%.

Starch Gelatinization

Native starches are present in semicrystalline granular forms with densities ca. 1.5 g/cm^3 [5]. Granular starch can be solvated to dispersed and amorphous molecules by heating with plasticizers (mainly water), or by treating them with organic solvents (e.g., dimethylsulfoxide), aqueous alkaline [7], or salt solutions (e.g., $CaCl_2$, KI) [8, 9]. The conversion from crystalline, granular starch to the dispersed and amorphous state is known as gelatinization [1].

Starch Retrogradation and Crystalline Structures

Solvated, amorphous starch has a tendency to return to an insoluble, aggregated, or crystalline state when being stored at a temperature above its glass transition temperature. This is known as retrogradation [1]. Retrogradation of starch dispersions is enhanced by a low temperature (0–5°C) and by high starch concentrations. Linear molecular chains have an extremely high tendency to retrograde. The retrogradation rate of linear molecules also depends on the chain length. Studies have shown that amylose with a chain length of ca. 100 glucose units has the greatest tendency to retrograde [10]. Retrograded branched molecules (amylopectin) can be more easily reversed than highly crystallized amylose. The latter requires high temperature (ca. 150°C) processing or chemical treatments (e.g., 90% DMSO, 0.5 N KOH).

X-ray crystallography of retrograded starch has revealed a double helical structure with a 10.5 Å repeat distance, indicating six glucose units per pitch (21 Å) [11]. In this structure, two parallel strands of starch are intertwined and the helix is stabilized by hydrogen bonds between the hydroxy groups and by hydrophobic interactions between the hydrocarbon moieties of the two strands. The double-helical crystalline starch is resistant to acidic and enzymatic hydrolyses. 55% of retrograded amylose sustained acid hydrolysis in 16% H_2SO_4 at 25°C for 40 days.

The resistant amylose consists of fragments of rather narrow molecular size distribution with the peak chain-length of ca. 31 glucose units [12].

Chemical Structures of Starch

Chemical structures of starch, such as molecular sizes of amylose, branch-chain lengths of amylopectin [13], and the proportion of amylose and amylopectin, are found to affect the functional properties of starch [13, 14]. Amylopectin of long branch-chains interacts more effectively with amylose and, thus, generates greater viscosity and gel strength. When the amylose content of the starch increases, the viscosity and gel strength of the starch paste also increase. Amylose alone and high-amylose maize starch produce strong films [14, 15].

In addition to amylose and amylopectin, starch also contains lipids (up to 1%), residues of protein (ca. 0.4%), and trace amounts of phosphorus (up to 0.09%) [16]. Lipids, particularly phospholipids, have great tendencies to form helical complexes with starch (mainly with amylose). Amylose–phospholipid complexes are so tightly bound that the lipids cannot be completely removed after 24 hours extraction with water-saturated butanol or 75% propanol solution in a Soxhlet extractor [16]. The amylose–lipid complex restricts the swelling of starch granules during cooking and results in a paste of reduced viscosity and increased opaque appearance. An example is wheat starch. Wheat-starch paste has significantly less viscosity than other starches because of its high phospholipid content.

Phosphate monoester derivatives are commonly found in starch. Potato starch is known for its great phosphate monoester content. A method to determine and differentiate phosphorous of different chemical structures in starch by using ^{31}P-NMR was developed [16]. The ^{31}P-NMR studies showed that most tuber and root starches, such as potato, sweet potato, arrowroot, and many legume starches, such as mung bean and green pea, have substantial amounts of phosphate monoester derivatives. Cereal starches, such as wheat, maize, and oat, have little or none [16]. Phosphate monoesters, carrying multiple negative charges which repel one another, enhance starch gelatinization and dispersion and retard starch retrogradation. The charged groups also have enhanced ionic interactions with proteins and other cationic groups. Starches with phosphate monoester derivatives, such as potato starch, produce pastes of increased clarity and viscosity. The paste also displays enhanced stability and a slow retrogradation rate.

Starch Morphology

Starches isolated from different botanical sources differ in their chemical structures [2] and in their morphologies [17]. Some starch granules, such as amaranth and pigweed, have submicron sizes, whereas others, such as canna and potato, have diameters of some granules larger than 100 μm. Many starches, such as potato and canna, are of oval shape; maize, oats, and rice starches are of polygonal and round shapes; wheat and barley starches are disk shaped and have bimodally distributed granular sizes. Some granules display unusual shapes, for example: diffenbachia starch is extended rod; shoti starch is a point-edged disk; babassu starch is half-spherical; all the legume starches investigated have beanlike granules each with a characteristic indent; and some high-amylose maize starch granules have budlike protrusions, whereas others are snake-like.

STARCH MODIFICATIONS AND APPLICATIONS

Starch, as an abundant and renewable resource for biodegradable polymers, is environmentally friendly and cost competitive. Properties of native starches may not be desirable for all applications. Many approaches, however, are available to modify the properties of starch to meet specific applications. These approaches include physical, chemical, and enzymatic modifications.

Physical Modifications of Starch

Starch can be physically modified to improve its water solubility and to change particle size. The requirement of heating or chemical treatment to disperse granules may be a hurdle for some applications. For example, the dispersion of high-amylose maize starch requires an extremely high processing temperature (ca. 150°C), which may require a specific equipment design and is not desirable for other ingredients in the mixture. Some active ingredients in starch mixtures are heat labile or volatile and may be destroyed by heating processes intended to disperse starch. In these cases, cold-water soluble starches would be desirable alternatives. Cold-water soluble starches can be prepared by various physical treatments.

Cold-Water-Soluble Starch

Pregelatinized starches can be prepared by precooking and drum-drying or by extrusion [18, 19]. Pregelatinized starch, after being rehydrated, displays a reduced viscosity, less sheen, and grainier paste than a freshly dispersed starch paste, indicating that some degradation and reassociation have taken place during processing.

To improve the viscosity and other physical properties, many technologies have been recently developed to produce granular cold-water-soluble (GCWS) starches. GCWS starch can be prepared by an injection and nozzle-spray drying process [20], high-temperature treatments of starch aqueous alcohol suspensions [21, 22], heating starch in aqueous polyhydric alcohol solutions [23, 24], and by alcoholic-alkali treatments [25, 26]. All these methods can be applied to normal starches. The alcoholic-alkali treatment method has a broader substrate spectrum, which can be used to prepare GCWS waxy as well as GCWS high-amylose starches [25]. The alcoholic-alkali method, however, cannot be applied to chemically-modified starches because most of them are labile at alkaline pH. Heating starch in an aqueous polyhydric alcohol solution has produced highly soluble GCWS chemically-modified starches, such as hydroxypropylated starch [23]. GCWS starches prepared by these methods display up to 96% water solubility at 25°C.

GCWS starches have been recently investigated as controlled-release agents for agriculture chemicals [27, 28]. GCWS waxy, normal, high-amylose V (50% amylose), and high-amylose VII (70% amylose) maize starches prepared by the alcoholic-alkali method were used to encapsulate atrazine for controlled-release studies [27]. Scanning electron micrographs show that atrazine is physically embedded in the starch matrix. Studies indicate that the swellibility of the matrix varies with the amylose content: the greater the amylose content, the less swelling of the matrix. The rate constant of atrazine release is inversely proportional to the amylose content of starch. These results indicate that starches of different amylose contents can be used to control chemical release rates [27].

Small-Particle Starch

Small-granular starch is desirable for use as plastic film fillers [29], cosmetic products, and fat replacements. Native small-granule starch is scarce and inherently expensive. A method has been developed to make small-particle starch from maize or from other abundant starches [30]. The method involves an acid treatment of granular starch to break glycosidic bonds in the amorphous region of the granule. The crystalline regions in the granule remain intact. After mechanical attrition of the acid-treated starch, the granule breaks up into small particles. Particle size of the products can be controlled by the degree of acid hydrolysis. The particle size can be reduced to submicron diameters if so desired.

Chemical Modifications of Starch

Chemical modifications of starch can be conducted to change starch properties [31]. Commonly used chemical modifications include crosslinking to enhance molecular stability against mechanical shearing, and acidic and high-temperature hydrolysis [31]. Hydroxyethylate, hydroxypropylate, phosphorylate, acetate, and succinate derivatives of starch can reduce the retrogradation rate [31]. Starch phosphate and succinate also have enhanced interaction with cationic molecules, whereas cationic starches have better interactions with electronegative substances, such as cellulose. Oxidized and partially hydrolyzed starches display decreased viscosities. Oxidized starches carry carbonyl and carboxyl groups, which also have enhanced interactions with cations or by crosslinks. Octenylsuccinate starch aluminum complex presents a hydrophobic surface of starch granules [31] and, thus, interacts more with hydrophobic matrices. Dialdehyde starch carries two aldehyde groups on each glucose unit [32]. The starch can develop a highly crosslinked network with other molecules through crosslinking reactions with $-OH$, $-NH_2$, and $-SH$ groups.

Starch maleate, octenylsuccinate, succinate, and dialdehyde starch have been investigated for starch-protein-based biodegradable plastics [33]. Starch maleate significantly enhances tensile strength and storage stability of molded plastic specimens [33]. Starch octenylsuccinate, however, increases the percentage elongation [33]. Dialdehyde starches enhance tensile strength and water resistance of molded plastic specimens [34]. Plastic specimens made from polymeric dialdehyde starch (90% oxidation) and zein (4:1/w:w) displays ca. 50 MPa and 2.7% water absorption after 24 hours submersion in water at 25°C [34].

Enzymatic Modifications of Starch

Starch can be enzymatically modified to produce maltodextrins, cyclodextrins, and oligosaccharides. Maltodextrins, prepared by α-amylase and/or acidic hydrolysis, have widely distributed molecular sizes. The smaller the molecular size, the lower the glass-transition temperature [35, 36]. Maltodextrins have been investigated as plasticizers to reduce the glass-transition temperature of materials [33]. Cyclodextrins have different molecular sizes (e.g., six, seven, and eight-member rings) and can be prepared by enzymes such as that from *B. macerans* [37]. Cyclodextrins consist of donutlike structures which have hydrophobic cavities. The hydrophobic cavity can complex with chemicals carrying hydrophobic moieties. Cylcodextrins have been used to stabilize aroma compounds during extrusion and thermal processing. It is not known how cyclodextrins affect the physical properties of plastics.

CONCLUSIONS

Naturally available starches have diverse properties which may meet different application requirements. Physical, chemical, and enzymatic modifications can tailor the properties of native starch for specific applications, otherwise requiring exotic starches.

REFERENCES

[1]   D. French, in *MTP International Review of Science: Biochemistry of Carbohydrates* (Biochemistry Series One, Vol. 5, W. J. Whelan, Ed.), Butterworths, London, 1975, p. 269.
[2]   D. R. Lineback, *Baker's Dig.*, p. 17 (1984).
[3]   K. Kainuma and D. French, *Biopolymers, 11*, 2241 (1972).
[4]   D. French, in *Starch: Chemistry and Technology*, 2nd ed. (R. L. Whistler, Ed.), Academic Press, Orlando, Florida, 1984, p. 184.
[5]   H. H. Wu and A. Sarko, *Staerke, 30*, 73 (1978).
[6]   S. Hizukuri, Y. Takeda, and M. Yasuda, *Carbohydr. Res., 94*, 205 (1981).
[7]   B. J. Oosten, *Starch/Staerke, 34*, 233 (1982).
[8]   J. Jane, *Ibid., 45*, 161 (1993).
[9]   I. Linqvist, *Ibid., 31*, 195 (1979).
[10] B. Pfannemuller, *Ibid., 38*, 401 (1986).
[11] A. D. French and V. G. Murphy, *Cereal Foods World, 22*, 61 (1977).
[12] J. Jane and J. F. Robyt, *Carbohydr. Res., 132*, 105 (1984).
[13] J. Jane and J. F. Chen, *Cereal Chem., 69*, 60 (1992).
[14] A. H. Young, in *Starch: Chemistry and Technology*, 2nd ed. (R. L. Whistler, Ed.), Academic Press, Orlando, Florida, 1984, p. 249.
[15] I. A. Wolff, H. A. Davis, J. E. Cluskey, L. J. Gundrum, and C. E. Rist, *Ind. Eng. Chem., 43*, 915 (1951).
[16] S.-T. Lim, T. Kasemsuwan, and J. Jane, *Cereal Chem., 71*, 488 (1994).
[17] J. Jane, T. Kasemsuwan, S. Lees, H. F. Zobel, and J. F. Robyt, *Starch/Staerke, 46*, 121 (1994).
[18] E. L. Powell, in *Starch: Chemistry and Technology*, Vol. II (R. L. Whistler and E. F. Paschall, Eds.), Academic Press, New York, 1967, p. 523.
[19] C. Mercier and P. Feillet, *Cereal Chem., 52*, 283 (1973).
[20] E. Pitchon, J. D. O'Rourke, and T. H. Joseph, US Patent 4,280,851 (1981).

[21]   J. E. Eastman and C. O. Moore, US Patent 4,465,702 (1984).

[22]   J. Jane, S. A. S. Craig, P. A. Seib, and R. C. Hoseney, *Starch/Staerke, 38*, 258 (1986).

[23]   S. Rajagopalan and P. A. Seib, *J. Food Sci., 16*, 13 (1992).

[24]   S. Rajagopalan and P. A. Seib, *Ibid., 16*, 29 (1992).

[25]   J. Chen and J. Jane, *Cereal Chem., 71*, 618 (1994).

[26]   J. Chen and J. Jane, *Ibid., 71*, 623 (1994).

[27]   J. Chen and J. Jane, *Ibid.*, In Press.

[28]   D. Trimnell and B. S. Shasha, *J. Control. Rel., 12*, 251 (1990).

[29]   S. Lim, J. Jane, S. Rajagopalan, and P. A. Seib, *Biotechnol. Prog., 8*, 51 (1992).

[30]   J. Jane, L. Shen, L. Wang, and C. C. Maningat, *Cereal Chem., 69*, 280 (1992).

[31]   O. B. Wurzburg, *Modified Starches: Properties and Uses*, CRC Press, Boca Raton, Florida, 1986, Chapter 8, p. 113.

[32]   C. L. Mehltretter, in *Methods in Carbohydrate Chemistry*, Vol. 4 (R. L. Whistler, Ed.), Academic Press, Orlando, Florida, 1964, p. 316.

[33]   S.-T. Lim and J. Jane, *J. Environ. Polym. Degrad., 2* (1994).

[34]   K. E. Spence, J. Jane, and A. L. Pometto III, *Ibid.*, In Press.

[35]   Y.-J. Wang and J. Jane, *Cereal Chem., 71*, 527 (1994).

[36]   H. Levine and L. Slade, *Carbohydr. Polym., 6*, 213 (1986).

[37]   K. Kainuma, in *Starch: Chemistry and Technology*, 2nd ed. (R. L. Whistler, Ed.), Academic Press, Orlando, Florida, 1984, p. 125.

MOLECULAR WEIGHT OF POLY(3-HYDROXYBUTYRATE) DURING BIOLOGICAL POLYMERIZATION IN *ALCALIGENES EUTROPHUS*

FUMITAKE KOIZUMI,† HIDEKI ABE, and YOSHIHARU DOI

Polymer Chemistry Laboratory
The Institute of Physical and Chemical Research (RIKEN)
Hirosawa, Wako-shi, Saitama 351-01, Japan

ABSTRACT

Poly(3-hydroxybutyrate) [P(3HB)] was produced by *Alcaligenes eutrophus* from fructose or butyric acid in nitrogen-free media at different pHs and temperatures, and the time-dependent changes in the molecular weight of P(3HB) were studied. P(3HB) polymers were accumulated within cells in the presence of a carbon source, and the rate of P(3HB) accumulation increased with temperature in the 16–35°C range but was almost independent of the pH value (5.0–8.0) of the media. The apparent activation energy for the rate of P(3HB) accumulation was 30 kJ/mol. The number-average molecular weight ($\overline{M}_n$) of P(3HB) polymers increased rapidly with time during the course of P(3HB) accumulation to reach a maximum value (($10 \pm 2 \times 10^5$)) at around 6 hours of incubation, followed by a gradual decrease in $\overline{M}_n$ with time. The polydispersities ($\overline{M}_w/\overline{M}_n$) of P(3HB) polymers increased from 1.5 to 2.0 during P(3HB) accumulation. The time-dependent changes in $\overline{M}_n$ have been accounted for in terms of the kinetic model in which a chain transfer agent is formed during the course of P(3HB) synthesis and reacts with a growing chain of P(3HB) on the active site of polymerase to regulate the length of P(3HB) chain.

†Current address: Tokyo Institute of Technology, Nagatsuta 4259, Midori-ku, Yokohama 227, Japan.

INTRODUCTION

A wide variety of bacteria synthesize an optically active polymer of (R)-3-hydroxybutyric acid (3HB) as an intracellular storage polymer, and P(3HB) is accumulated as granules within the cytoplasm [1–4]. *Alcaligenes eutrophus* used in this study accumulates P(3HB) in the cells in amounts up to 70% of the dry weight, when growth is limited by the depletion of an essential nutrient but the cells have an excess of carbon source [5]. The structure of native P(3HB) granules in *A. eutrophus* cells was characterized by means of ^{13}C-NMR spectroscopy [6] and x-ray diffraction [7], which demonstrated that the P(3HB) granules were completely amorphous. The P(3HB) polymer can be extracted from bacterial cells with a suitable organic solvent such as chloroform or methylene chloride. The isolated P(3HB) is a partially crystalline thermoplastic with biodegradable and biocompatible properties [3].

The pathway and regulation of P(3HB) synthesis have been studied extensively in *A. eutrophus* [8–11]. P(3HB) is synthesized from acetyl-coenzyme A (CoA) by a sequence of three enzymatic reactions. 3-Ketothiolase catalyzes the reversible condensation reaction of two acetyl-CoA molecules into acetoacetyl-CoA, the intermediate is reduced to (R)-3-hydroxybutyryl-CoA by NADPH-linked acetoacetyl-CoA reductase, and P(3HB) is then produced by the polymerization of (R)-3-hydroxybutyryl-CoA with P(3HB) polymerase (synthase). The DNA sequences of the genes encoding these enzymes have been analyzed [12–14]. Recent immunocytochemical analysis in *A. eutrophus* revealed the localization of P(3HB) polymerase at the surface of P(3HB) granules [15]. The degradation of P(3HB) is initiated by P(3HB) depolymerase to form (R)-3-hydroxybutyric acid [16]. In a previous paper [17] we suggested that the intracellular P(3HB) depolymerase is an *exo*-type hydrolase which hydrolyzes an ester bond of P(3HB) at the terminus of a polymer chain.

The time-dependent changes in the molecular weights of P(3HB) polymers during the course of P(3HB) synthesis have been studied in *A. eutrophus* by several groups [17–20]. The number-average molecular weights ($\overline{M}_n$) of P(3HB) polymers were found to decrease with time during the synthesis of P(3HB). However, little is known about the regulation in molecular weight of P(3HB) polymers within bacterial cells. In a previous paper [17] we studied the kinetics of synthesis and degradation of P(3HB) in *A. eutrophus* and proposed a kinetic model involving chain propagation and chain transfer reactions of P(3HB) on the active site of polymerase.

In this paper we investigate time-dependent changes in the molecular weight of P(3HB) polymers produced in *A. eutrophus* from fructose or butyric acid under various conditions and extend the kinetic model of biological polymerization.

EXPERIMENTAL SECTION

Alcaligenes eutrophus H16 (ATCC 17699) was used in this study. The cell growth and P(3HB) accumulation within cells was carried out on a two-stage batch cultivation of *A. eutrophus*. The first stage was for cell growth. *A. eutrophus* cells were grown under aerobic conditions at 30°C for 24 hours in a nutrient-rich medium

(1.0 L) containing 10 g polypetone, 10 g yeast extract, 5 g meat extract, and 5 g $(NH_4)_2SO_4$. The cells were harvested by centrifugation at 4500g for 10 minutes at 4°C. Under these growth conditions, P(3HB) was not accumulated in the cells. The second stage was for P(3HB) accumulation. About 4g (dry weight) of the collected cells without P(3HB) was transferred into a nitrogen-free mineral medium (1 L) containing different amounts of Na_2HPO_4 and KH_2PO_4, 0.2 g $MgSO_4$, and 1 mL microelement solution [21]. To adjust the pH value of the mineral medium, different amount of 0.5 M Na_2HPO_4 (A) and 0.5 M KH_2PO_4 (B) were mixed in the medium (1 L) as listed below. When necessary, the pH value of the medium was adjusted by the addition of 1N NaOH and H_2SO_4 solution.

		pH		
	5.0	5.8	7.0	8.0
(A) mL	159	159	39.0	7.38
(B) mL	13.8	13.8	53.6	64.2

Different amounts of fructose and butyric acid were added to the media as the sole carbon source. The cells were aerobically incubated at different temperatures from 16 to 35°C in a 2.6-L jar fermenter equipped with six conventional turbine impellers and three baffles.

Each 20 mL of culture was collected periodically during the incubation. The supernatant was removed by centrifugation at 4500g for 10 minutes at 4°C, and the collected cells were washed with distilled water and lyophilized. P(3HB) was extracted from the lyophilized cells into chloroform by stirring for 48 hours at room temperature, and the cell material was removed by filtration. P(3HB) samples were purified by reprecipitation with hexane and dried in vacuo.

The concentrations of fructose in nitrogen-free media were determined at 25°C by the UV-method with a biochemical analysis kit (Boehringer Mannheim GMBH Biochemica) on a Hitachi U-2000 spectrophotometer. The concentration of butyric acid was determined by gas chromatography analysis as described in a previous paper [21] and by dissolved organic carbon (DOC) analysis using a Shimadzu TOC-5000 analyzer.

To determine the P(3HB) content in bacterial cells, 8–10 mg lyophilized cells was reacted with a solution containing 2 mL chloroform, 1.7 mL methanol, and 0.3 mL sulfuric acid for 4 hours at 100°C. After the reaction, 1 mL distilled water was added and the tube was shaken for 1 minute. After phase separation the organic phase was removed and used for GC analysis. The resulting methyl esters in the organic phase were measured with n-octanoic acid methyl ester as a standard by gas chromatography using a Shimadzu GC-14A with a Neutra Bond-1 column and a flame ionization detector.

Molecular weight data of P(3HB) samples were obtained at 40°C by gel permeation chromatography (GPC) using a Shimadzu LC-9A system equipped with Shodex 80M and K-802 columns and RID-6A refractive index detector. Chloroform was used as the eluent at a flow rate of 0.8 mL/min, and a sample concentration of 1 mg/mL was used. The number-average and weight-average molecular weights

$(\overline{M}_n$ and $\overline{M}_w)$ were calculated by a Shimadzu Chromatopac C-R4A with a GPC program. A molecular weight calibration curve of P(3HB) was obtained on the basis of the universal calibration method [22] with polystyrene standards of low polydispersities.

RESULTS

The effect of the initial concentration (5–20 g/L) of the carbon source (fructose or butyric acid) on the rate of P(3HB) accumulation by *A. eutrophus* was reported in a previous paper [17]. The rates of fructose consumption and of P(3HB) accumulation were independent of the concentration of fructose. In contrast, a high concentration (20 g/L) of butyric acid caused a decrease in the rates of butyrate consumption and of P(3HB) accumulation. In this study the effects of pH and temperature on the rate of P(3HB) accumulation were studied in nitrogen-free media of *A. eutrophus*. A two-stage batch fermentation was used for the kinetic analysis of P(3HB) accumulation in *A. eutrophus* cells. In the second stage, no cell growth took place in nitrogen-free media, and P(3HB) polymers accumulated within cells.

P(3HB) Production from Fructose

Figure 1 shows the time courses of P(3HB) accumulation and degradation in *A. eutrophus* cells in the nitrogen-free media containing 20 g/L fructose as the sole carbon source at 30°C. The pH values of the media remained constant at 5.0, 5.8, 7.0, and 8.0, respectively, during the incubation. The concentration of fructose in the media decreased linearly with time and reached zero at around 30 hours of incubation (Fig. 1C). In contrast, the amounts of P(3HB) in cells increased proportionally to time while fructose was present in the medium (Fig. 1B). These results suggest that the concentration of monomer ((R)-3-hydroxybutyryl-CoA) in cells remains constant during the synthesis of P(3HB). After the carbon source was exhausted, P(3HB) gradually degraded with time, indicating that P(3HB) was utilized for energy generation under conditions of carbon starvation.

Figure 2 (A and B) shows the time-dependent changes in the number-average molecular weight ($\overline{M}_n$) and polydispersities ($\overline{M}_w/\overline{M}_n$) of P(3HB) produced from fructose at different pH values. Figure 2(C) shows the time-dependent changes in the number of P(3HB) polymer chains ($[N]$) during incubation. The number of P(3HB) polymer chains at time t ($[N]_t$) can be calculated by

$$[N]_t = Y_t/\overline{M}_{n,t} \tag{1}$$

where Y_t and $\overline{M}_{n,t}$ are the yield and number-average molecular weight of P(3HB) at time t, respectively. The $\overline{M}_n$ value of P(3HB) increased rapidly with time and reached a maximum value (800,000–1,100,000) at around 6 hours of incubation, followed by a gradual decrease in $\overline{M}_n$ during the course of P(3HB) accumulation (Fig. 2A). The polydispersities ($\overline{M}_w/\overline{M}_n$) of P(3HB) decreased from 2.2 ± 0.3 to 1.6 ± 0.2 during the initial stage of incubation for 4 hours, followed by a gradual increase to 2.0 ± 0.3 with time (Fig. 2B). On the other hand, the $[N]$ value increased with time during the course of P(3HB) accumulation (Fig. 2C).

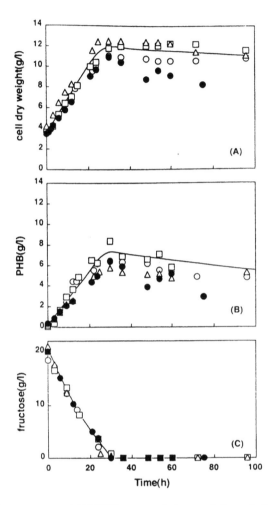

FIG. 1. Time courses of P(3HB) accumulation and degradation during the batch
ition of *A. eutrophus* in nitrogen-free media (1 L) containing fructose (20 g/L) as a
irbon source at 30°C. The pH values of media remained constant at (○) 5.0, (□) 5.8,
0, and (●) 8.0.

Table 1 lists the rate of P(3HB) accumulation (R_p), the rate of P(3HB)
dation (R_d), and the rate of increase in the number of P(3HB) polymer chains
at different pH values of the nitrogen-free media (Runs 1, 2, 3, and 4) as
nined from Figs. 1(B) and 2(C). The rate of P(3HB) accumulation was
it independent of the pH value of the medium, but the rate of increase in the
er of P(3HB) polymer chains decreased slightly with an increase in pH value.
Figure 3 shows the time courses of P(3HB) accumulation and degradation in
trophus cells in nitrogen-free media containing 20 g/L fructose at pH 7.0. The
:ratures of the media were kept constant at 16, 23, 27, 30, and 35°C, respec-
, during the incubation. The rates of P(3HB) accumulation and fructose con-
tion increased with temperature. The maximum weights of P(3HB) accumu-
in cells are listed in Table 1. The maximum weights (7.0 ± 1.3 g/L) of

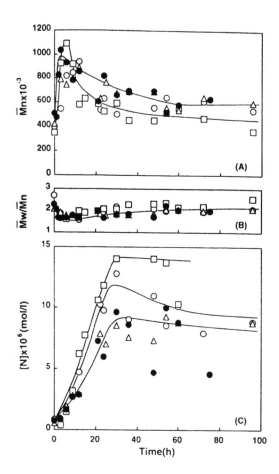

FIG. 2.   Time-dependent changes in the number-average molecular weights ($\overline{M}_n$) (A), polydispersities ($\overline{M}_w/\overline{M}_n$) (B), and the chain number ([N]) (C) of P(3HB) polymers produced from fructose in nitrogen-free media (1 L) at pH values of ($\bigcirc$) 5.0, ($\square$) 5.8, ($\triangle$) 7.0, and ($\bullet$) 8.0.

P(3HB) in *A. eutrophus* cells were not influences by pH and the temperature of the media. Thus, the yields of P(3HB) from fructose were 35 ± 6 wt%, independent of incubation conditions.

Figure 4 (A and B) shows the time-dependent changes in the values of $\overline{M}_n$ and $\overline{M}_w/\overline{M}_n$ of P(3HB) produced from fructose at different temperatures. The $\overline{M}_n$ values increased rapidly with time to reach a maximum value (900,000–1,400,000) at around 6 hours of incubation, followed by a gradual decrease in $\overline{M}_n$ with time. The values of $\overline{M}_w/\overline{M}_n$ decreased with time to reach a minimum value (1.5 ± 0.2) at around 6 hours of incubation, followed by a gradual increase to 1.9 ± 0.2 with time. Figure 4(C) shows the time-dependent change in the number of P(3HB) polymer chains ([N]) during incubation at different temperatures.

Table 1 gives the rates of R_p, R_d, and R_N at different temperatures, determined from Figs. 3(B) and 4(C). The temperature dependences of R_p and R_N are plotted in Fig. 5. The apparent activation energies of the rate of P(3HB) accumulation (R_p)

TABLE 1. Maximum Weight of P(3HB), Content of P(3HB) in Dry Cell, Rate (R_p) of P(3HB) Accumulation, Rate (R_d) of P(3HB) Degradation, and Rate (R_N) of Increase in the Number of P(3HB) Polymer Chains under Different Incubation Conditions in Nitrogen-Free Media Containing Fructose or Butyric Acid as the Sole Carbon Source for A. eutrophus

Run	Concentration of substrate, (g/L)	pH	Temperature, °C	Maximum weight, g/L	Content, wt%	$R_p \times 10^3$, mol/h·L	$R_d \times 10^4$, mol/h·L	$R_N \times 10^7$, mol/h·L
Fructose								
1	20	5.0	30	6.4	58	2.6 ± 0.3	3.3 ± 0.4	4.3 ± 0.3
2	20	5.8	30	8.3	73	3.3 ± 0.5	7.1 ± 0.8	5.5 ± 1.0
3	20	7.0	30	5.7	46	2.3 ± 0.5	3.2 ± 0.2	3.2 ± 0.3
4	20	8.0	30	6.3	58	2.3 ± 0.3	7.7 ± 0.6	3.0 ± 0.4
5	20	7.0	16	7.3	72	1.5 ± 0.3	n.d.	2.1 ± 0.5
6	20	7.0	23	6.2	52	1.7 ± 0.3	1.6 ± 0.2	2.0 ± 0.1
7	20	7.0	27	7.4	62	2.8 ± 0.5	4.4 ± 1.0	2.9 ± 0.2
8	20	7.0	35	7.2	60	3.2 ± 0.5	5.7 ± 0.2	3.0 ± 0.8
Butyric acid								
9	10	7.0	20	3.5	42	0.82 ± 0.2	3.7 ± 0.5	2.1 ± 0.2
10	10	7.0	25	4.2	50	2.7 ± 0.4	6.9 ± 0.2	4.1 ± 0.5
11	10	7.0	30	4.3	55	3.2 ± 0.5	3.2 ± 0.6	4.5 ± 1.0
12	10	7.0	35	4.2	47	4.0 ± 0.2	4.9 ± 0.5	6.0 ± 1.5

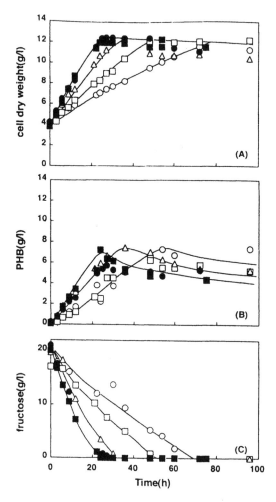

FIG. 3.   Time courses of P(3HB) accumulation and degradation during the batch incubation of *A. eutrophus* in nitrogen-free media containing fructose (20 g/L) as a sole carbon source at pH 7.0. The temperatures of media remained constant at (○) 16, (□) 23, (△) 27, (●) 30, and (■) 35°C.

and of the increase in the number of P(3HB) polymer chains (R_N) were determined as 30 and 20 kJ/mol, respectively.

P(3HB) Production from Butyric Acid

In this experiment, butyric acid was used as the sole carbon source for *A. eutrophus* in nitrogen-free media. Figure 6 shows the time courses of P(3HB) accumulation and degradation in *A. eutrophus* cells in the nitrogen-free media containing 10 g/L butyric acid at pH 7.0. The temperature of the medium remained constant at 20, 25, 30, and 35°C, respectively, during the incubation. The rates of P(3HB) accumulation and butyrate consumption increased with temperature. The maximum weights of P(3HB) accumulated in cells are given in Table 1. The values

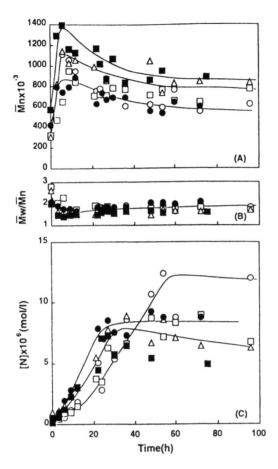

FIG. 4.   Time-dependent changes in the number-average molecular weights ($\overline{M}_n$) (A), polydispersities ($\overline{M}_w/\overline{M}_n$) (B), and the chain number ($[N]$) (C) of P(3HB) polymers produced from fructose in nitrogen-free media (1 L) at temperatures of (○) 16, (□) 23, (△) 27, (●) 30, and (■) 35°C.

(4.2 ± 0.1 g/L) were independent of temperature in the 25–35°C range. The yields of P(3HB) from butyric acid were 42 wt% at temperatures above 25°C, while the yield at 20°C was 35 wt%.

Figure 7 shows the time-dependent changes in $\overline{M}_n$ (A), $\overline{M}_w/\overline{M}_n$ (B), and the number of polymer chains $[N]$ (C) of P(3HB) produced from butyric acid at different temperatures. The $\overline{M}_n$ values increased with time to reach a maximum value at 5 hours and then decreased to a constant value with time. It is important to note that the $\overline{M}_n$ values of P(3HB) remain almost constant during the degradation of P(3HB) in the absence of butyric acid. This result suggests that a gradual decrease in $\overline{M}_n$ during the course of P(3HB) accumulation is not related to the degradation of P(3HB) by the intracellular depolymerase. The $\overline{M}_w/\overline{M}_n$ values increased to a constant value (2.0 ± 0.3) with time. The $[N]$ values increased with time during the course of P(3HB) accumulation in the presence of butyric acid. The rates of increase in the $[N]$ value in the initial stage for 5 hours were slower than the rates

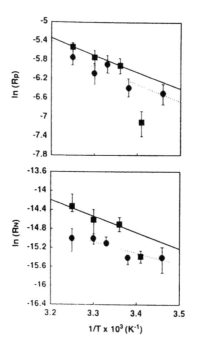

FIG. 5.   Arrhenius plots of ln R_p (A) and ln R_N (B) against $1/T$. The filled squares (■) and circles (●) are the data obtained from butyric acid and fructose, respectively.

after 5 hours. After the carbon source was exhausted, the [N] values decreased with time. The rates of decrease in [N] values in Fig. 7(C) are higher than the rates in Fig. 4(C).

Table 1 lists the rate of P(3HB) accumulation (R_p), the rate of P(3HB) degradation (R_d), and the rate of increase in the [N] value (R_N) at different temperatures, determined from Figs. 6(B) and 7(C). The temperature dependence of R_p and R_N are shown in Fig. 5. The apparent activation energies of R_p and R_N were determined as 30 and 28 kJ/mol, respectively. Thus, the activation energy of P(3HB) accumulation by *A. eutrophus* from butyric acid was consistent with the value (30 kJ/mol) from fructose.

DISCUSSION

In a previous paper [17] we proposed the reaction scheme of initiation, chain propagation, and chain transfer in the enzymatic polymerization of (R)-3-hydroxybutyryl-CoA with P(3HB) polymerase as follows.

Initiation:

$$\text{E-SH} + (R)\text{-CH}_3\text{CH(OH)CH}_2\text{CO-SCoA} \xrightarrow{k_i}$$

$$(R)\text{-CH}_3\text{CO(OH)CH}_2\text{CO-S-E} + \text{CoA-SH} \tag{2}$$

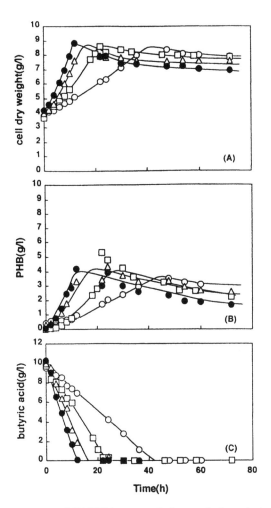

FIG. 6.   Time courses of P(3HB) accumulation and degradation during the batch incubation of *A. eutrophus* in nitrogen-free media (1 L) containing butyric acid (10 g/L) as a carbon source at pH 7.0. The temperatures of media remained constant at ($\bigcirc$) 20, ($\square$) 25, ($\triangle$) 30, and ($\bullet$) 35°C.

Chain propagation:

$$P(3HB)_{n-1}OCH(CH_3)CH_2CO\text{-}S\text{-}E + (R)\text{-}CH_3CH(OH)CH_2CO\text{-}SCoA$$

$$\xrightarrow{\ k_p\ } \ P(3HB)_nOCH(CH_3)CH_2CO\text{-}S\text{-}E + CoA\text{-}SH \qquad (3)$$

Chain transfer:

$$P(3HB)_{n-1}OCH(CH_3)CH_2CO\text{-}S\text{-}E + X(H_2O) \ \xrightarrow{\ k_t\ }$$

$$P(3HB)_{n-1}OCH(CH_3)CH_2COOH + E\text{-}SH + X \qquad (4)$$

where E-SH, CoA-SH, (R)-CH$_3$CH(OH)CH$_2$CO-SCoA, P(3HB)$_n$, and X denote P(3HB) polymerase with thiol groups as the active site, coenzyme A, (R)-3-

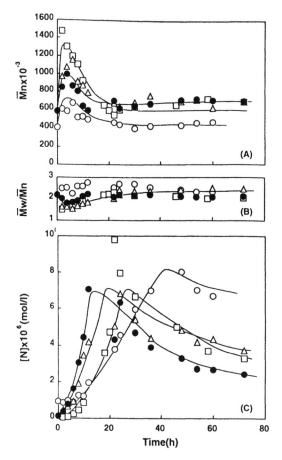

FIG. 7.   Time-dependent changes in the number-average molecular weights ($\overline{M}_n$) (A), polydispersities ($\overline{M}_w/\overline{M}_n$) (B), and the chain number ([N]) (C) of P(3HB) polymers produced from butyric acid in nitrogen-free media (1 L) at temperatures of ($\bigcirc$) 20, ($\square$) 25, ($\triangle$) 30, and ($\bullet$) 35°C.

hydroxybutyryl-CoA as monomer, a polymer chain with an n polymerization degree of 3HB units, and a chain-transfer agent, respectively, and k_i, k_p, and k_t are the pseudo-first-order rate constants for initiation, propagation, and chain-transfer reactions. The chain transfer agent X may bind a water molecule as an active form.

In this paper we extend our kinetic model proposed in a previous paper [17]. A key enzyme of P(3HB) synthesis is P(3HB) polymerase with thiol groups as the active site [11, 13, 23]. The polymerase active of *A. eutrophus* was found to remain almost constant during P(3HB) accumulation [11]. This result suggests that the number of polymerase molecules remains constant during P(3HB) accumulation. As shown in Figs. 2, 4, and 7, the number of P(3HB) polymer chains [N] increased with time during the accumulation of P(3HB) with an induction period of several hours, and the polydispersities ($\overline{M}_w/\overline{M}_n$) of molecular weight distribution of P(3HB) polymers increased to 2.0 with time. These results indicate that a chain-transfer reaction of the propagating chain takes place on the P(3HB) polymerase.

As a result of the chain-transfer reaction, the length of the growing P(3HB) chain is limited, and the $[N]$ value increases with time.

In this study an induction period with an increase in the number of P(3HB) polymer chains $[N]$ was observed for several hours in the initial stage of P(3HB) accumulation (see Figs. 2, 4, and 7), which suggests that the rate of chain-transfer reaction increases with time in the initial stage of biological polymerization. Here we propose a reaction scheme in which the concentration of a chain-transfer agent (X) increases with time to reach a constant value, resulting in an increase in the rate of chain transfer. A chain-transfer agent (X) may be formed from an inactive compound A after the synthesis of P(3HB) is initiated within cells as,

$$A \xrightarrow{a} X \quad (d[X]/dt = a[A]) \tag{5}$$

where a is the rate constant of the formation of chain-transfer agent X. The concentration of chain-transfer agent X at time t is represented by

$$[X] = [X]_0(1 - e^{-at}) \tag{6}$$

with $[X]_0 = [A] + [X]$. Then the rate of chain-transfer reaction (R_t) at time t is given by

$$R_t = k_t[E][X] = k_t[E][X]_0(1 - e^{-at}) \tag{7}$$

where $[E]$ is the concentration of P(3HB) polymerase molecules and k_t is the rate constant of the chain-transfer reaction. The number of P(3HB) polymer chains $[N]_t$ at time t is given by Eq. (8) from Eq. (7):

$$[N]_t = [E] + \int_0^t R_t dt = [E] + k_t[E][X]_0\{t + (e^{-at} - 1)/a\} \tag{8}$$

TABLE 2.   Rate Constants of Chain Propagation (k_p), Chain Transfer Reaction $(k_t[X]_0)$, and Formation of a Chain Transfer Agent (a)

Run	k_p, h^{-1}	$k_t[X]_0$, h^{-1}	a, h^{-1}
1	5200	0.86	0.20
2	6600	1.1	0.20
3	4600	0.64	0.20
4	4600	0.60	0.23
5	3000	0.40	0.11
6	3400	0.40	0.20
7	5600	0.58	0.25
8	6400	0.60	0.21
9	1600	0.42	0.09
10	5400	0.82	0.13
11	6400	0.90	0.19
12	8000	1.2	0.30

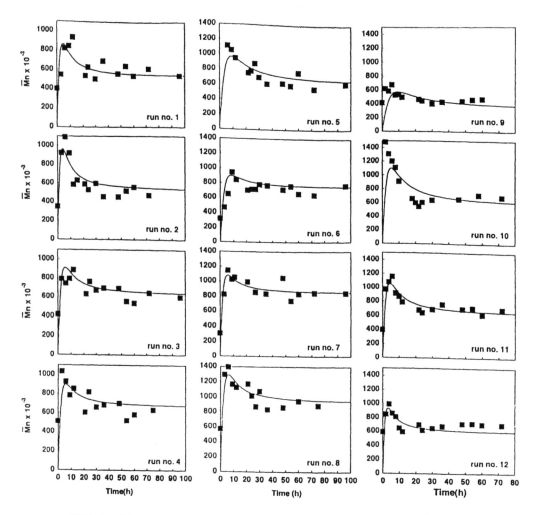

FIG. 8.   Time courses of the number-average molecular weights ($\overline{M}_n$) of P(3HB) polymers obtained in Runs 1 to 12. The experimental data are shown by filled squares, and the values calculated by Eq. (11) with the rate constants in Table 2 are represented by solid lines.

Here we make the assumptions that 1) there is a rapid initiation reaction and 2) the number of polymerase molecules [E] remains constant during P(3HB) synthesis. From Eq. (8), we have $[N] = [E]$ at $t = 0$ and $[N] = k_t[E][X]_0 t$ at $t = \infty$. The value [E] may be determined by extrapolation of $[N]_t$ values to time zero, using the plots of $[N]_t$ vs t in Figs. 2, 4, and 7. It was estimated that the number of polymerase molecules was $(5 \pm 2) \times 10^{-7}$ mol/L. The value of [E] is the same value as reported in a previous paper [17]. The values of $k_t[X]_0$ and of the rate constant a for the formation of chain-transfer agent X were determined from the values of R_N (Table 1) and the relations of $[N]_t$ vs t plots in Figs. 2, 4, and 7 by using Eq. (8). The values of $k_t[X]_0$ and a are given in Table 2.

    The yield Y_t of P(3HB) polymers produced for time t is given by

$$Y_t = \int_0^t R_p dt = k_p[E]t \tag{9}$$

where R_p and k_p denote the rate of chain propagation and the pseudo-first-order rate constant for propagation, respectively. The values of k_p were determined from the values of R_p (Table 1) and the [E] value (5×10^{-7} mol/L), and they are listed in Table 2.

The number-average polymerization degree ($\overline{P}_{n,t}$) of P(3HB) polymers produced for time t is given by Eq. (10):

$$\overline{P}_{n,t} = \frac{Y}{[N]_t} = \frac{\int_0^t R_p dt}{[E] + \int_0^t R_t dt}$$

$$= \frac{k_p[E]t}{[E] + k_t[E][X]_0\{t + (e^{-at} - 1)/a\}} \tag{10}$$

The rate constants k_p and k_t are assumed to be constant during polymerization. Then, the number-average molecular weight ($\overline{M}_{n,t}$) of P(3HB) polymers produced in time t is given by

$$\overline{M}_{n,t} = \frac{86k_p[E]t}{[E] + k_t[E][X]_0\{t + (e^{-at} - 1)/a\}} \tag{11}$$

where 86 is the molecular weight of a 3HB unit.

The $\overline{M}_{n,t}$ values at time t were calculated from Eq. (11) by using the values of k_p, $k_t[X]_0$, and a in Table 2. The results are shown in Fig. 8. The experimental data of $\overline{M}_n$ for Runs 1 to 12 are given by closed squares, and the calculated values of $\overline{M}_n$ are represented by solid lines. The calculated values of $\overline{M}_n$ are in good agreement with the experimental data for all runs. Thus, an unusual time-dependent change in $\overline{M}_n$ during the synthesis of P(3HB) may be interpreted in terms of the model that a chain-transfer agent X is formed from an inactive compound A during the synthesis of P(3HB). The chain-transfer agent X may be a coenzyme. More research is needed to identify the chain-transfer agent in this biological polymerization.

This study was supported by the Special Coordination Research Fund of the Science and Technology Agency, Japan.

REFERENCES

[1] E. A. Dawes and P. J. Senior, *Adv. Microbiol. Physiol.*, *10*, 135 (1973).
[2] A. J. Anderson and E. A. Dawes, *Microbiol. Rev.*, *54* 450 (1990).
[3] Y. Doi, *Microbial Polyesters*, VCH Publishers, New York, 1990.

[4]   A. Steinbüchel, in *Biomaterials* (D. Byrom, Ed.), Macmillan, London, 1991, Chapter 3.

[5]   H. G. Schlegel, G. Gottschalk, and R. V. Bartha, *Nature (London), 191,* 463 (1961).

[6]   G. N. Barnard and J. K. M. Sanders, *J. Biol. Chem., 264,* 3286 (1989).

[7]   Y. Kawaguchi and Y. Doi, *FEMS Microbiol. Lett., 70,* 151 (1990).

[8]   V. Odeing and H. G. Schlegel, *Biochem. J., 134,* 239 (1973).

[9]   G. W. Haywood, A. J. Anderson, L. Chu, and E. A. Dawes, *FEMS Microbiol. Lett., 52,* 91 (1988).

[10] G. W. Haywood, A. J. Anderson, L. Chu, and E. A. Dawes, *Ibid., 52,* 259 (1988).

[11] G. W. Haywood, A. J. Anderson, and E. A. Dawes, *Ibid., 57,* 1 (1989).

[12] O. P. Peoples and A. J. Sinskey, *J. Biol. Chem., 264,* 15293 (1989).

[13] O. P. Peoples and A. J. Sinskey, *Ibid., 264,* 15298. (1989).

[14] P. Schubert, N. Krüger, and A. Steinbüchel, *J. Bacteriol., 173,* 168 (1991).

[15] T. U. Gerngross, P. Railly, J. Stubbe, A. J. Sinskey, and O. P. Peoples, *Ibid., 175,* 5289 (1993).

[16] H. Hippe and H. G. Schlegel, *Arch. Mikrobiol., 56,* 278 (1967).

[17] Y. Kawaguchi and Y. Doi, *Macromolecules, 25,* 2324 (1992).

[18] D. G. H. Ballard, P. A. Holmes, and P. J. Senior, in *Recent Advances in Mechanistic and Synthesis Aspects of Polymerization* (M. Fonille and A. Guyot, Eds.), Reidel, Lancaster, UK, 1987, p. 219.

[19] R. Bradel, A. Kleinke, and K. H. Reichert, *Makromol. Chem., Rapid Commun., 12,* 583 (1991).

[20] R. Bradel and K. H. Reichert, *Makromol. Chem., 194,* 1983 (1993).

[21] Y. Doi, Y. Kawaguchi, Y. Nakamura, and M. Kunioka, *Appl. Environ. Microbiol., 55,* 2932 (1989).

[22] Z. Grubisic, R. Rempp, and H. Benoit, *J. Polym. Sci., Part B, 7,* 3676 (1968).

[23] A. Steinbüchel, E. Hustede, M. Liebergesell, U. Pieper, A. Timm, and H. Valentin, *FEMS Microbiol. Rev., 103,* 217 (1992).

BIODEGRADABLE BLENDS OF CELLULOSE ACETATE AND STARCH: PRODUCTION AND PROPERTIES

JEAN M. MAYER

US Army Natick RD&E Center
Natick, Massachusetts 01760-5020

GLENN R. ELION

International Communications and Energy
Chatham, Massachusetts 02633

CHARLES M. BUCHANAN

Eastman Chemical Company
Kingsport, Tennessee 37622

BARBARA K. SULLIVAN and SHELDON D. PRATT

University of Rhode Island
Narragansett, Rhode Island 02882-1197

DAVID L. KAPLAN

US Army Natick RD&E Center
Natick, Massachusetts 01760-5020

ABSTRACT

Blends of cellulose acetate (2.5 degree of substitution) and starch were melt processed and evaluated for mechanical properties, biodegradability during composting, and marine and soil toxicity. Formulations containing, on a weight basis, 57% cellulose acetate (CA), 25% corn starch (St) and 19% propylene glycol (PG) had mechanical properties

183

similar to polystyrene. Increasing plasticizer or starch content lowered tensile strength. Simulated municipal composting of cellulose acetate alone showed losses of 2–3 and 90% dry weight after 30 and 90 days, respectively. CA/St/PG blends in both soil burial and composting experiments indicate that propylene glycol and starch are degraded first. Extended incubations are required to detect losses from cellulose acetate. Marine toxicity tests using polychaete worms and mussels showed no toxicity of cellulose acetate or starch. High doses had an adverse effect due to oxygen depletion in the marine water due to rapid biodegradation of the polymers. Preliminary plant toxicity tests of the CA/St blends showed no negative impact on growth and yield for sweet corn, butternut squash, and plum tomatoes. The results indicate that CA/St blends have acceptable properties for injection-molded applications and are biodegradable and nontoxic.

INTRODUCTION

Concerns over entrapment and ingestion hazards associated with persistent plastics in the environment have spurred research in developing materials which can function like plastics during storage and use, yet are broken down into nontoxic by-products if disposed in the environment. Many approaches have been taken to commercially produce either biodegradable or photodegradable plastics [1]. One approach involves the incorporation of starch into blends with synthetically produced polymers which are made to be susceptible to biological, chemical, or UV breakdown. Alternatively, fermentation has been used to convert sugars or waste biomass into polyesters such as polyhydroxybutyrate/valerate or lactic acid for polylactic acid production which have been shown to be biodegradable and/or hydrolyzable. Limitations in producing and commercializing these biodegradable formulations are production costs, which can be 2–10 times higher than commodity plastics such as polystyrene or polyethylene, and lack of biodegradation and toxicity test standards for intended disposal environments.

Cellulose acetate is a commodity plastic produced from cotton linters or wood pulp and used as textile fibers, injection-molded items, and coatings. Its performance characteristics and biodegradability have been linked to the degree of substitution (DS). Studies conducted in simulated aerobic sludge [2] and composting [3, 4] have shown that cellulose acetate with a DS of 3 was not biodegradable, while at 2.5 DS the polymer was slowly degraded and at DS <2.2 was readily biodegraded. Since the cost of cellulose acetate is approximately three times higher than that of polystyrene, blending with inexpensive plasticizers or fillers is desirable to lower product cost without severely compromising mechanical properties. In this paper we discuss the effects of blending starch and propylene glycol with cellulose acetate on mechanical properties and biodegradability. Marine and soil toxicity testing of cellulose acetate and starch is also presented since marine disposal and composting are potential disposal routes for these materials.

EXPERIMENTAL

Materials Preparation

Cellulose acetate/starch blends were mixed, extruded, and pelletized as previously described [5]. Cellulose acetate with a DS of 2.5 (CA 398-30, Eastman Chemical, Kingsport, Tennessee) was mixed with a variety of corn starches, Argo brand (Best Foods, Englewood Cliffs, New Jersey), Melogel (National Starch and Chemical Co., Bridgewater, New Jersey), and Crisp-Tex and Amalean I (American Maize Products Co., Hammond, Indiana), or potato (Difco Laboratories, Detroit, Michigan) or rice starch (California Products, San Diego, California) and food grade propylene glycol (Dow Chemical Corp., Midland, Michigan) in a Henschel brand high intensity mixer (Purnell International, Houston, Texas) at 3000 rpm for several minutes. Calcium carbonate (Omyacarb UF, Omya Products, Lucerne Valley, California) was added to some samples to neutralize residual acid. The $CaCO_3$ was first suspended in the propylene glycol at low mixing speed prior to the addition of starch and cellulose acetate.

The powdered mixture was fed into a 42-mm diameter twin screw extruder (C. W. Brabender Instruments, Inc., Hackensack, New Jersey) equipped with a four hold (3.2 mm) strand die. The screw speed was 60 rpm and the temperature profile was as follows: zone 1, 120°C; zone 2, 150°C; zone 3, 160°C; and zone 4 (die), 170°C. The strands were then pelletized and the pellets heated at 185°C and molded into 38 mm long × 3.2 mm thick dogbone tensile bars in a labtop scale injection molder (Custom Scientific Instruments, Cedar Knolls, New Jersey).

Physical Testing

Mechanical Properties.   Tensile strength and modulus of the dogbone samples were determined using a Model 4204 tensile strength tester (Instron Inc., Canton, Massachusetts) using a crosshead speed of 1.3 mm/min.

Capillary Rheometry.   High impact polystyrene (Dow Chemical Corp., Midland, Michigan) and the blend of cellulose acetate 398-30 (57% wt/wt), cornstarch (25% wt/wt), and propylene glycol (18% wt/wt) pellets were analyzed in a capillary rheometer (Instron Inc., Canton, Massachusetts) at shear rates between 20 and 2000 (1/s) at temperatures between 170 and 225°C.

Scanning Electron Microscopy

Cellulose acetate 398-30, starches, and cellulose acetate/starch blend morphologies were examined using a S-900 Field Emission Gun SEM (Hitachi Instruments, Mountain View, California). Samples were coated with 50 Å tungsten and viewed under low voltage (1 keV).

Biodegradation Testing

Composting Studies.   Cellulose acetate (398-30, 2.5 DS) films and cellulose acetate/starch blend dogbones were exposed in simulated municipal solid waste composting reactors as described previously [6]. The compost mixture had the following composition: 3500 g dehydrated alfalfa meal, 1300 g cottonseed meal, 1400 g poplar sawdust, 1000 g fresh cow manure, 1500 g black garden soil, 2500 g finely shredded newspaper, 480 g $CaCO_3$, 40 g $NaHCO_3$, and 13 L water, which resulted in a C:N ratio of 30:1. After blending in a Hobart Mixer to obtain a 3–4 mm particle size, the mixture was placed in a 276 mm (o.d.) by 432 mm length stainless steel cylinder with screens on the ends and airtight seals for aeration or gas sampling. The natural microorganisms found in the compost ingredients served as the inoculum. Samples were incubated for 15 to 90 days, and weight loss was determined after washing and drying the samples to a constant weight.

Soil Burial. The laboratory soil burial system consisted of 1 part commercial sand to 1 part topsoil to 1 part composted cow manure (1881 Select brand, Earth Grow Inc., Lebanon, Connecticut) containing 30% by weight moisture [7]. Cellulose acetate/starch dogbones were placed in the soil mixture and incubated at 30°C. Samples were periodically removed after 14 to 112 days, washed, dried to constant weight, and the percent weight loss and weight loss/surface area determined.

Toxicity Testing

Marine Mussel Toxicity. *Mytilus edulis* were grown in large mesocosms (1.8 m diameter × 5.5 m height tanks) which were filled with marine water containing natural populations of organisms from Narragansett Bay, Rhode Island, to which 10 mg/L starch or cellulose acetate was added. After exposure for 3 weeks in sample and control tanks, growth of the mussels was measured, and mussel meat and fecal material were analyzed for differences in stable isotope values [8].

Benthic Core Toxicity. Benthic cores containing natural populations from Narragansett Bay were collected and supplemented with 10 and 50 mg/L starch and cellulose acetate. The number of viable species, mostly polychaete worms, were enumerated in the control and test cores before and after a 1-month incubation.

Plant Toxicity Testing. Cellulose acetate (57% wt/wt)/starch (25% wt/wt)/ propylene glycol (19% wt/wt) pellets were powdered in a disk mill until a particle size of approximately 300 μm was produced. A 1% (wt/wt) loading of the powdered blend was then added to potting soil and placed in greenhouse pots. Twenty pots, each containing four seeds, were used for both control and test experiments. Sweet corn, butternut squash, and plum tomatoes were used as test plants in the study. When the control plants reached maturation, all of the plants were harvested and measured for plant length and fruit weight. The percent change in size/weight compared to the control was then determined.

RESULTS AND DISCUSSION

Physical Properties

Initially, cellulose acetate was blended with propylene glycol alone to determine the plasticizer range required to produce injection-molded items which have high strength without being too brittle. Figure 1 shows the tensile strength at break and Young's modulus for injection-molded dogbones made from powdered samples or pellets produced from the powders by twin screw extrusion. Strength and modu-

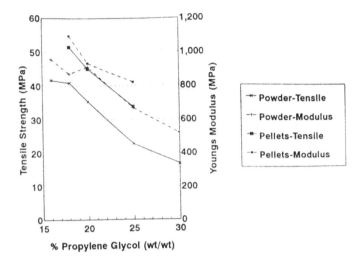

FIG. 1. Mechanical properties of cellulose acetate/propylene glycol tensile bars produced from powders and extruded pellets ($N = 5$).

lus increase with decreasing plasticizer content, and a 16–20% propylene glycol range results in better properties. Tensile strength values show the differences between the powdered and pelletized samples at different plasticizer contents more consistently than the modulus.

The effect of starch type on the tensile strength of injection-molded dogbones with 57% cellulose acetate and 19% propylene glycol is shown in Fig. 2. Corn starch performed better in the blends than potato or rice starch. Among the corn starches, the Melogel and Argo brands, which contain 30:70 amylose:amylopectin, performed similarly. The two blends containing 50:50 amylose:amylopectin brands,

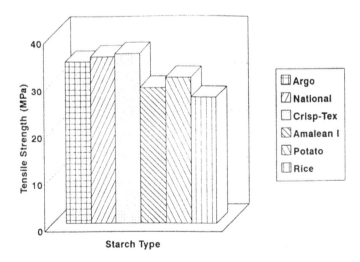

FIG. 2. Effect of starch type on tensile properties ($N = 5$).

TABLE 1.   Effect of $CaCO_3$ on Mechanical Properties of Cellulose
Acetate/Argo Starch/Propylene Glycol (60:20:17 wt/wt) Blends

Formulation	Tensile strength (MPa)	Young's modulus (MPa)
Polystyrene	22.1	905
CA/St	35.6	924
CA/St 2.5% $CaCO_3$	40.4	1060
CA/St 5% $CaCO_3$	36.2	974

Crisp-Tex and Amalean I, had very different strength properties which could be
attributed to differences in compatibility with cellulose acetate. Crisp-Tex has a low
degree of acetylation whereas Amalean I is derivatized with a low DS of propylene
oxide.

   Pellets produced from cellulose acetate, starch, and propylene glycol had an
acidic smell and taste. Therefore, calcium carbonate ($CaCO_3$) was added to the
blend in an attempt to neutralize residual acetic acid released during extrusion.
Table 1 shows that addition of 2.5 and 5.0% $CaCO_3$ improves mechanical properties
and also eliminated the taste and odor problems. The dogbones molded from CA/
St/PG/$CaCO_3$ blends had better tensile strength and modulus than polystyrene
(Table 1). However, in performance tests with utensils molded from this blend, the
impact resistance was not as high as for polystyrene (data not shown).

   Capillary rheometry studies were performed to compare flow, at different
temperatures and shear rates, of the CA/St blends to high impact polystyrene to
facilitate production of injection-molded utensils. The results presented in Fig. 3.
indicate that the viscosity curve for the CA/St blend is similar to polystyrene. These
results guided the successful production of prototype utensils during the first trial
run.

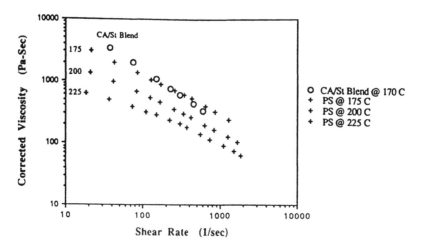

FIG. 3.   Comparison of the rheology of cellulose acetate/Argo starch blends with
polystyrene.

Scanning Electron Microscopy

Electron micrographs of corn and potato starch before and after blending and extrusion with cellulose acetate are presented in Fig. 4. The corn starch granules appear to be unbroken within the cellulose acetate matrix; however, the potato starch granules look partially disrupted. There is evidence of void formation between the starch granules and cellulose acetate in some areas, and some coating of CA on the granules.

Biodegradation Testing

Figure 5 shows the weight loss/surface area data for CA/St blends in compost and soil burial exposures and cellulose acetate alone in composting. The cellulose acetate film lost 50 μm/mm^2 surface area (>90% total weight) during the 90-day test period. The blends (in tensile bars) in soil and compost studies lost approximately 20% of their total weight (approximately 230 to 270 μg/mm^2), which is probably indicative of leaching or degradation of the propylene glycol and partial leaching or biodegradation of the starch. We would expect the organisms present in compost or soil to attack the most readily degradable substrates first.

Toxicity

Polymers can be transformed into metabolites that have greater toxicity due to the action of microorganisms. While it is not expected that biodegradable polymers will form toxic by-products, testing should be done to verify their benign nature.

Mussel Toxicity. Filter feeders such as mussels are sensitive to pollutants in the marine environment. Stable carbon isotope ratio of the mussel biomass and fecal material, if different from the ratio in cellulose acetate and starch, can be used to determine if these materials are ingested and metabolized by the mussel, ingested but excreted without being metabolized, or not ingested. When related to changes in growth, reproduction, or mortality, this technique can help to pinpoint the mode of toxicity. It has been successfully used in tracing the toxicological fate of sewage sludge in sediments and estuarine ecosystems upon ingestion by phytoplankton and benthic organisms [8]. Large mesocosms, which are filled with marine water and sediment containing natural fish larvae populations, have been used to mimic an ecosystem response to exposure to starch-containing biodegradable plastics [9]. The results from these experiments showed that larval fish growth rate was reduced because they ingested the starch blend but could not digest it. In the experiment with mussels, growth was not inhibited by the presence of starch or cellulose acetate in the mesocosms. Table 2 indicates that starch was ingested based on the change in the fecal stable carbon isotope value, but neither starch nor cellulose acetate was incorporated into mussel biomass.

Benthic Core Toxicity. Figure 6 shows the effects of addition of starch or cellulose acetate to sediment cores containing benthic organisms, primarily polychaete worms. The number of viable species recovered was reduced at dose levels of

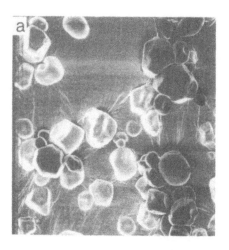

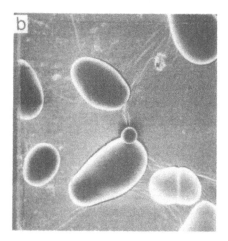

FIG. 4.   Scanning electron micrographs of (a) corn starch granules, (b) potato starch granules, (c) cellulose acetate/corn starch pellets, and (d) cellulose acetate/potato starch pellets (1000×).

TABLE 2.   Stable Carbon Isotope ($^{12}C/^{13}C$) Ratios of Mussel Meat and Feces after Exposure to Starch and Cellulose Acetate 398–30

Sample	Carbon isotope, initial	Carbon isotope in meat, 3 weeks	Carbon isotope in feces, 3 weeks
Mussel control	−19.2	−19.0	−14.1
CA-398-30	−31.7	−18.8	No data
Starch	−10.1	−18.9	−11.5

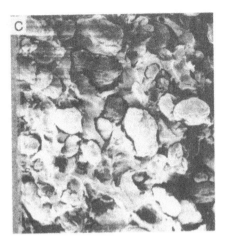

FIG. 4 (continued).

50 mg/L, particularly with the starch-treated cores. The appearance and smell of hydrogen sulfide in these cores indicates that the problem is a secondary one, the reduction in oxygen, and not direct toxicity of the rapidly biodegraded starch.

Plant Toxicity Testing.   The primary route being considered for disposal of biodegradable materials is composting. An important consideration for this disposal route is verification that there is no plant toxicity from the compost containing the biodegradable polymers, formulation additives, or their breakdown products. The effect of addition of CA/St blends on the growth and fruit production of corn, squash, and tomatoes when directly applied in the potting soil was evaluated (Table 3). The blend slightly improves the growth and yield of these vegetables.

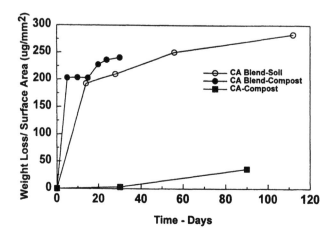

FIG. 5.   Weight loss/surface area of cellulose acetate (2.5 DS) in compost and cellulose acetate/corn starch/propylene glycol blends in compost and soil.

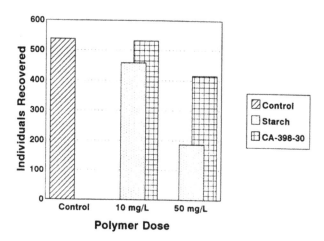

FIG. 6.   Effect of starch and cellulose acetate on the mortality of benthic organisms.

TABLE 3.   The Effect of CA/St Blend on
Plant Growth and Yield

Plant	Plant length (% control)	Yield weight (% control)
Corn	105	105
Squash	110	115
Tomato	110	115

CONCLUSIONS

Cellulose acetate/starch blends were produced which process similarly to polystyrene and have similar mechanical properties. The type and amount of starch as well as plasticizer content influenced mechanical performance. Cellulose acetate with a DS of 2.5 was shown to be slowly biodegraded in compost, and addition of starch and propylene glycol seems to delay the utilization of the CA portion of the blend. Marine and plant toxicity tests indicate that these blends should be safe if disposed of in the marine environment or through composting.

ACKNOWLEDGMENTS

We wish to thank Betty Ann Welsh and Renay Pollier for assisting with processing, Paul Dell for conducting the capillary rheometry work, Deborah Vezie for taking the SEM photographs, and Robert Stote and Joseph McCassie for conducting the soil biodegradation studies.

REFERENCES

[1] L. McCarthy-Bates, *Plas. World*, p. 22 (1993).
[2] C. M. Buchanan, R. M. Gardner, and R. J. Komarek, *J. Appl. Polym. Sci.*, *47*, 1709 (1993).
[3] C. M. Buchanan, R. M. Gardner, R. J. Komarek, S. C. Gedon, and A. W. White, *Biodegradable Polymers and Packaging*, Technomic, Lancaster, Pennsylvania, 1993, p. 133.
[4] R. A. Gross, J. D. Gu, D. T. Eberiel, M. Nelson, and S. P. McCarthy, *Ibid.*, p. 257.
[5] J. M. Mayer and G. R. Elion, US Patent 5,288,318 (February 22 1994).
[6] R. M. Gardner, C. M. Buchanan, R. J. Komarek, D. Dorchel, C. Boggs, and A. W. White, *J. Appl. Polym. Sci.*, In Press.
[7] J. M. Mayer, M. Greenberger, and D. L. Kaplan, *Polym. Prepr., Div. Polym. Mater.: Sci Eng., ACS*, p. 63 (1990).
[8] P. J. Gearing, J. N. Gearing, J. T. Maughan, and C. A. Oviatt, *Environ. Sci. Technol.*, *25*(2), 295 (1991).
[9] G. Klein-MacPhee, B. K. Sullivan, and A. A. Keller, *Am. Fish. Soc. Symp.*, *14*, 105 (1993).

PRESENT AND FUTURE OF PLA POLYMERS

M. VERT, G. SCHWARCH, and J. COUDANE

University Montpellier 1
CRBA-URA CNRS 1465
Faculty of Pharmacy
15 ave. Charles Flahault, 34060 Montpellier, France

ABSTRACT

Lactic acid-based aliphatic polyesters (PLAs) are well known bio-compatible bioresorbable polymers which are being increasingly used as biomaterials for temporary therapeutic applications. Because of their sensitivity to water and the formation of degradation by-products which can be easily metabolized by microorganisms, this type of polymers also has potential to replace commodity polymers in packagings or as mulch films. From an overview of synthesis routes, structural characteristics, and performances, an attempt is made to evaluate the future of PLA polymers insofar as industrial development is concerned.

INTRODUCTION

High molecular weight aliphatic polyesters of the poly(2-hydroxy acid) type were synthesized for the first time more than 40 years ago when the ring-opening polymerization of 1,4-dioxane-2,6-dione (glycolide = GA) and 3,5-dimethyl-1,4-dioxane-2,6-dione (lactide = LA) was discovered [1, 2]. For many years these polymers, which form the PLA/GA family, were regarded as useless compounds because of their sensitivity to heat and water which precluded thermal processings based on extrusion or injection molding. This situation ended in the 1960s when advantage was taken of their sensitivity to water to make artificial degradable sutures of the poly(glycolic acid) type, PGA, aimed at replacing denatured collagen known as Catgut [3, 4]. In 1966, Kulkarni et al. [5] showed that high molecular weight poly(L-lactic acid) and poly(DL-lactic acid) were also degradable in vivo and

195

thus of interest for biomedical applications. Since then, poly(2-hydroxy acids) have been studied extensively with respect to potential applications as temporary therapeutic aids in surgery and in pharmacology [6]. PLA/GA devices for bone fracture internal fixation (Biofix from Finland, Phusilines from France) and antitumoral drug delivery systems (Zoladex from UK, Decapeptyl from France, Enanthone from Japan) are now commercially available. Because of their biocompatibility, their degradability, the mineralization or metabolization of their degradation by-products, and the fact that a large range of properties can be covered by taking advantage of copolymerization and stereocopolymerization, PLA/GA polymers are presently considered as the most attractive compounds for temporary therapeutic applications in the biomedical field [6].

During the last 5 years, people have realized that polymers were becoming a source of ecological problems because of the huge amounts of bioresistant wastes generated from the 70 millions tons of polymers produced each year in the world.

Beside more or less problematic solutions like incineration, reprocessing of wastes, pyrolysis, or other chemical degradations aimed at regenerating monomers, bioassimilation of polymers by environmental microorganisms after degradation or biodegradation is regarded with interest, at least from a basic viewpoint, because it could offer a solution by eliminating dispersed packaging wastes from the environment or mulch films from culture fields. They could also be used for pesticides, insecticides, or hormones in local delivery in plant therapy on the basis of the concept of sustained release drug delivery. Ideally, such applications should require polymers functional for the targeted application and also biocompatible, i.e., capable of respecting environmental life and allowing biological recycling if corresponding monomers are released from the biomass and degraded by-products can return to it.

In this paper we present an overview of the present knowledge on the origins, the structural characteristics, and the performance of lactic acid polymers in order to comment on their future with respect to industrial development. We will examine successively the situation insofar as synthesis, chirality, thermal properties, mechanical properties, and degradation mechanisms are concerned.

THE FAMILY OF LACTIC ACID-BASED POLYMERS

The literature is quite confusing about poly(lactic acid) polymers mostly because authors usually do not clearly define the type of PLA they are dealing with. This is primarily due to the fact that many authors belong to other disciplines and are thus not well informed about polymer science and the effects of chain composition and chain structure on the macroscopic properties of polymeric materials. In order to easily reflect the composition of LA-derived polymer chains, we have introduced simple acronyms, namely PLA_XGA_Y where X = percentage of L-LA units and Y = percentage of GA units [7]. Of course, these acronyms can be adjusted to include comonomeric units other than GA in the case of copolymers derived from other cyclic compounds such as caprolactone or ethylene oxide (Table 1). The use of these acronyms is not yet generalized in the literature.

TABLE 1. Chemical Structures and Acronyms of Poly(α-Hydroxy Acids) Derived from Lactic and Glycolic Acids

Chemical structure	Acronyms
$$-[O-\overset{\displaystyle H}{\underset{\displaystyle CH_3}{\overset{\displaystyle |*}{C}}}-CO-]_{\overline{n}}$$	PLA_{100} poly(L-lactic acid)
$$-[O-\overset{\displaystyle H}{\underset{\displaystyle CH_3}{\overset{\displaystyle |*}{C}}}-CO-\vdots_{\overline{n}}\text{comonomer } B-]_{\overline{q}}$$	$PLA_{(100-Y)}B_Y$ $(Y = q/(n + q))$ (L-LA/GA copolymers)
$$-[O-\overset{\displaystyle H}{\underset{\displaystyle CH_3}{\overset{\displaystyle |*}{C}}}-CO-\vdots_{\overline{n}}O-\overset{\displaystyle CH_3}{\underset{\displaystyle H}{\overset{\displaystyle |*}{C}}}-CO-]_{\overline{p}}$$	PLA_X $(X = 100n/(n + p))$ (LA stereocopolymers)
$$-[O-\overset{\displaystyle H}{\underset{\displaystyle CH_3}{\overset{\displaystyle |*}{C}}}-CO-\vdots_{\overline{n}}O-\overset{\displaystyle CH_3}{\underset{\displaystyle H}{\overset{\displaystyle |*}{C}}}-CO-\vdots_{\overline{p}}\text{ comonomer } B-]_{\overline{q}}$$	PLA_XB_Y $(X = 100n/(n + p + q))$ $(Y = 100q/(n + p + q))$ (terpolymers)

MONOMERIC SPECIES AND SYNTHESIS OF LACTIC ACID-BASED POLYMERS

Lactic acid-based polymers can be produced from various monomers and by different routes (Figs. 1 and 2). The poly(2-hydroxy acid)-type backbone can be formed by polycondensation or step-growth polymerization of lactic acid with other hydroxy acids (Fig. 2) [8–10]. It can also be obtained by chain-growth polymerization of lactide with other cyclic monomers such as caprolactone or ethylene oxide. Lactide presents the particularity of being composed of two lactic acid units linked by two ester bonds to form a dimeric cyclic monomer (Fig. 2). Both lactic acid and lactide are chiral, and thus one has to deal with two optical isomers or enantiomers in the case of lactic acid and four isomers, or diastereoisomers, in the case of lactide. This cyclic dimer bears two assymetric carbon atoms which can be identical, as in the case of L-lactide, D-lactide, and DL-lactide, or different as in the case of *meso*-lactide (Fig. 2). The latter is rather difficult to prepare in a high optical purity form and is rarely mentioned in the literature. L-Lactic acid is a product of biotechnologies. Racemic lactic acid can be obtained either by racemization of L-lactic acid or by synthesis through oil chemistry. L- and DL-Lactide are obtained in two steps from L-lactic acid and DL-lactic acid, respectively [11]. *meso*-Lactide is extracted from the filtrates remaining after DL-lactide recrystallization.

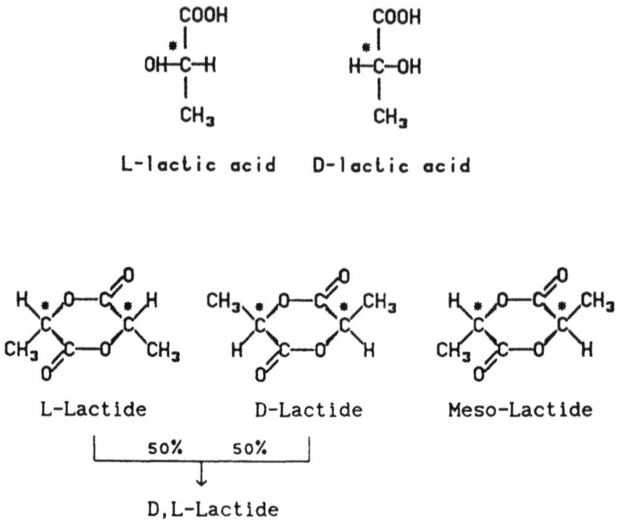

FIG. 1.   Lactic acid and lactide monomers.

The polycondensation of hydroxy acids is generally performed by distillation of water, with or without catalyst, vacuum and temperature being progressively increased [8–10]. This is a rather simple step which leads to oligomers with low molecular weights (< 5000 daltons, $M_w/M_n \approx 2$). Temperatures above 180°C usually cause coloration. Higher molecular weights (50,000) have been obtained by postcondensation in organic solution using dicyclohexylcarbodiimide (DCC) [12]. However, this method has not been developed industrially.

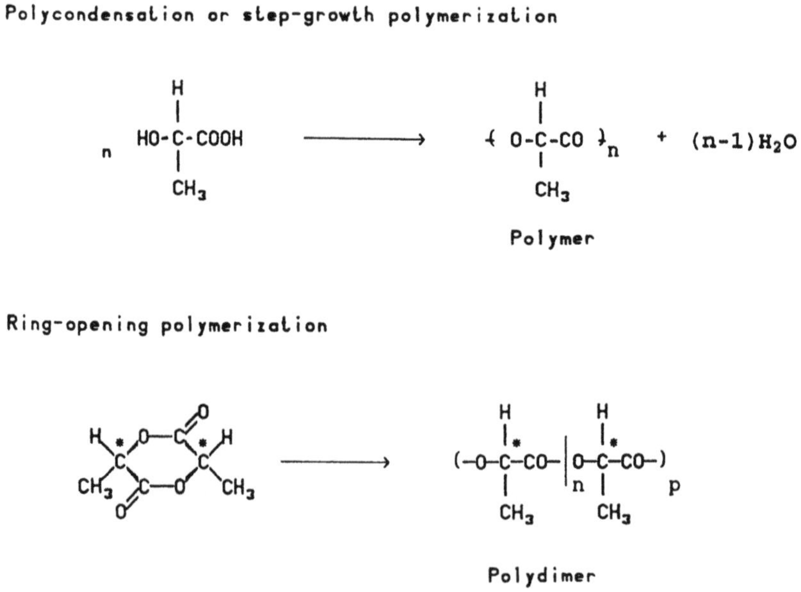

FIG. 2.   Routes to make poly(2-hydroxy acid)-type backbones.

So far, high molecular weight polymers are obtained from the ring-opening polymerization of lactide-containing feeds, either in the bulk at different temperatures or in organic solution [11]. Many compounds can initiate the ring-opening polymerization of lactide-type cyclic dilactones. In the case of lactide, the literature shows that almost all the possible mechanisms (anionic [13, 14], cationic [15–17], or by insertion after complexation [18–23]) have been reported. Critical factors are the initiator concentration [21, 23–26]; the polymerization temperature [16, 21, 23, 26, 27]; the polymerization time [13, 16, 21, 23–27]; the presence of acids [11], of water [23], or of alcohols used as coinitiators [28]; and other impurities. On the other hand, secondary reactions such as depolymerization with monomer regeneration at equilibrium [23, 25, 29, 30] or transesterification [13, 20, 21, 31] are generally observed and must be controlled. With respect to industrialization, the use of an organic solvent is possible; however, bulk polymerization would be preferred if the problem of heat transfer in large batches could be solved [11]. In the biomedical and pharmacological fields, stannous octoate is generally recommended for initiation because it gives rather fast polymerization and has been approved by the US Food and Drug Administration. Many years ago [32] we selected powdered zinc metal as a convenient initiator for biomedical applications because zinc is biocompatible and a useful nutritional oligoelement. This initiator requires longer polymerization times than stannous octoate, especially for L-lactide-rich feeds. It is now used by our group as a standard in order to make the comparisons of the data collected over the years more consistent. Combined with purification by the solvent/nonsolvent technique, zinc leads to polymers which are extrudable or processible by compression or injection moldings. Heat processed devices like bone plates, screws, and rods are now currently made industrially in different countries, at least on a small scale. Sometimes people do not purified the crude polymer and process it as-polymerized [33]. In this case, molding is impossible and the devices have to be machined. In any case, the lactide monomers have to be carefully dried to yield high molecular weight compounds, and processing has to be carried out carefully with polymers of intermediate molecular weight in order to preserve the quality of the initial polymer and to obtain devices with good mechanical properties. Films, micro-, and nanoparticulates can be made from solutions in organic solvents. Slight degradation is generally observed, but the main problem is the removal of residual solvents. This can be achieved at high temperatures and under vacuum, an operation which is not always acceptable because of the risk of deformation and degradation.

STRUCTURAL CHARACTERISTICS OF LACTIC ACID-BASED POLYMERS

As recalled above, the PLA family includes copolymers with other monomers and thus can present all the classical structures generally found in polymer science, namely random, diblock or multiblock, star, etc., with all the consequences one can predict for such differences. Molecular weight is also an important parameter, as for all polymers. However, the most striking feature is chirality. The presence of asymmetric carbon atoms generates structural particularities which make lactic acid-derived polymers rather special as compared with other polymers. For the sake of clarity, comments will be limited to homo- and stereocopolymers of lactic acid

(PLA$_X$) (Table 1). However, the distribution of the chiral repeating units can be very different, depending of the synthesis route and on the chiral monomer(s) [30, 34, 35]. Polycondensation of optically pure L-lactic acid and ring-opening polymerization of optically pure L-lactide should lead to the same isotactic chain structure, provided no racemization occurs during chain growth. It is of interest to note that minor differences can be generated when a fragment of the initiator remains anchored at one or two chain ends as in the case of the ring-opening polymerization via the insertion mechanism. The same is true for D-lactic acid and D-lactide which are mirror images of L-lactic acid and L-lactide, respectively. In contrast, dramatic differences in main chain structures can be observed as soon as one deals with stereocopolymers composed of L- and D-lactic acid repeating units. Step-growth polymerization of mixtures of L- and D-lactic acids leads to a random distribution of the L- and D-units, whereas ring-opening polymerization of the lactide dimer leads to nonrandom distributions because the chain grows through a pair addition mechanism. These structural differences were identified many years ago [34]. However, only a few authors have taken the corresponding consequences into account [30, 34–36]. A similar comment can be made in the case of transesterification reactions and racemization which both tend to cause chain randomization although total racemization is extremely rare.

PHYSICAL AND MECHANICAL PROPERTIES

Basically, lactic acid polyesters (PLA$_X$) have the characteristics of a glassy material with T_g in the 55–60°C range. Because of the presence of one chiral center per repeating unit, the corresponding macromolecules are more or less stereoregular, and thus the macroscopic morphology depends very much on X, with variations being similar in the symmetrical ranges $50 < X < 100$ and $50 > X > 0$. D-Units in a poly-L chain behave like impurities. However, melting temperature and crystallinity decrease rather rapidly with increasing X because of the pair addition mechanism. PLA 92 is still slightly crystalline whereas PLA 87.5 is amorphous. Highly crystalline PLA$_X$ can be obtained with high mechanical properties, especially Young's modulus and tensile strength. PLA 100 is a strong but brittle material. Less stereoregular PLA 98 and 96 have better mechanical characteristics due to their lower crystallinity but not to the plasticizing effect of D-units since T_g is almost the same for all PLA$_X$ under dry conditions [7]. Interestingly enough, oligomers can be used to plasticize the polymers, a feature of great interest for adjusting mechanical characteristics [37]. As for all polymers, the physical and mechanical characteristics depend very much on processing history, on molecular weight, and on molecular weight distribution. Because of the difficulty to control all these parameters, compounds mentioned in literature are far from being comparable, even if they bear the same name. This remark was made for the first time 13 years ago [7] and it is still valuable today. Finally, PLA macromolecules appear very versatile because they can lead from waxy amorphous oligomeric compounds to very strong crystalline devices. Carefully processed stereoregular PLA polymers are rather resistant to atmospheric moisture and aqueous media at normal temperatures. Highly crystalline PLA 100 can last more than 10 years under normal conditions of moisture and temperature, provided it has been carefully processed, but this is not true when impurities are present in the polymer matrix. Therefore, one must keep in mind that the degradability of lactic acid-based polymers depends on many more factors than was initially thought, as underlined a few years ago [38].

HYDROLYTIC DEGRADATION

The very first investigations revealed that PLA/GA polymers degrade hydrolytically and that the cleavage of ester bonds is catalyzed through carboxylic chain ends. The number of chain ends increases as degradation advances, the catalytic effect also increases, and hydrolytic degradation is said to be "autocatalytic" [39, 40]. A couple of years ago the phenomenon of heterogeneous degradation was discovered to be typical of the hydrolytic degradation of PLA/GA large-size, thick devices [41]. This phenomenon is characterized by a faster degradation inside than at the surface, leading, for some of the members of the PLA/GA family, to hollow structures (Fig. 3). This interpretation has been questioned regarding the possibility of a skin effect due to molding history [42]. Actually, autocatalysis can be the source of many other phenomena such as size-dependence of the degradation rate [43], nonrandom chain cleavages, and degradation-induced morphology changes [44]. For example, it has been recently shown that large size devices degrade much faster than submillimetric particles or very thin films [45]. In other words, the thinner the device, the slower the hydrolytic degradation. These features cannot be explained by molding heterogeneities. However, sooner or later (4 weeks up to several years, depending on many factors [38]), PLA/GA polymers will be returned to their constituting hydroxy acids and will be assimilated by living cells.

As for the ability of living cells or microorganisms to attack PLA/GA surfaces by an enzymatic process, no conclusive demonstration has been made available so far, although many authors have mentioned some modifications of degradation rates in enzyme-containing aqueous media on the basis of with/without experiments [45]. Presently it seems that only water-soluble oligomeric compounds released from partially degraded or oligomer-containing matrices can be degraded in the presence of cells as high molecular weight poly(hydroxybutyrate) does, for example [46]. Recently, a careful in-vitro investigation of the degradation of PLA stereoco-

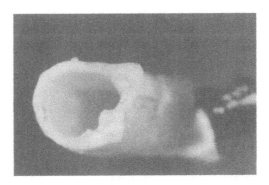

FIG. 3.   Part of hydrolytically degraded specimen showing the faster internal degradation (2 × 2 mm initially).

polymers in the presence of proteinase K has shown that solid PLA does biodegrade [47]. However, this work did not demonstrate that PLA is biodegradable in vivo or under real outdoor conditions. In this respect, PLA/GA polymers are presently degradable compounds which can be referred to as "bioresorbable" with respect to assimilation or excretion by animal bodies, and as "bioassimilable" with respect to the giant living system formed by the planet [48].

FUTURE OF LA-BASED POLYMERS

For many reasons, namely rather complex structures, complex degradation mechanisms, ignorance of polymer science, etc., a great deal of confusion exists in PLA/GA literature, generally because investigations have been carried out on poorly defined compounds. Many people talk about PLA as if it is a single magic compound. Actually, as for polystyrene, polyethylene, poly(vinyl chloride), and other commodity polymers, PLA reflects a large family of compounds whose future will depend very much on structure–property relationships. In this respect, degradable LA-based polymers offer a large range of characteristics which have to be confronted in the lists of specifications of various time-limited applications one can find in human activities. The field of biomedical and pharmacological applications is now opened, and markets are becoming available which can absorb rather high added values. Plant therapy and related problems might be potential markets, but research is only starting in this field. So far, LA-based polymers are still expensive materials if one refers to packaging applications. However, chemical engineers have not yet exhausted all the possible solutions to the problem of cheap monomer and polymer productions. A United States company, Cargil, recently claimed a continuous process to make high molecular weight PLAs cheaply. Insofar as cost-effectiveness is concerned, one can predict that production costs will certainly never be as low as are those of present biostable polymers. Processing LA-based polymers is always a delicate operation. However, it seems that special machines with careful control of humidity and temperature might permit processing of large amounts of polymers. Therefore, the future seems to be mostly dependent on first defining what a life-respecting polymer has to be. We are still totally ignorant about the possible impacts of degradation by-products on living and nonliving environmental systems. A great deal of research will be necessary in order to fill the void. If there are finally practical niches, the question will be the creation of new markets, and the future will depend primarily on the attitude of society and consumers with respect to extra costs. From a technical viewpoint, LA-based polymers appear as good candidates for further industrial applications if bioassimilation becomes a strict prerequisite in the field of life-respecting materials. There is no doubt that the intensity of current research efforts on PLA/GA polymers will bring further insights on the perspectives and thus more accurate and consistent predictions.

REFERENCES

[1]   C. E. Lowe, US Patent 2,668,162 (1954).
[2]   A. K. Schneider, US Patent 2,703,316.
[3]   E. E. Schmitt and R. A. Polistina, US Patent 3,463,158 (1969).

[4]   E. J. Frazza and E. E. Schmitt, *J. Biomed. Mater. Res. Symp., 1*, 43 (1971).
[5]   R. K. Kulkarni, K. C. Pani, C. Neuman, and F. Leonard, *Arch. Surg., 93*, 839 (1966).
[6]   M. Vert, *Angew. Makromol. Chem., 166/167*, 155 (1989).
[7]   M. Vert, F. Chabot, J. Leray, and P. Christel, *Makromol. Chem., Suppl. 5*, 30 (1981).
[8]   J. Kleine, British Patent 779,291 (1957).
[9]   H. Fukuzaki, M. Yoshida, M. Asano, and M. Kumakura, *Eur. Polym. J., 25*, 1019 (1989).
[10] H. Fukuzaki, M. Yoshida, M. Asano, and M. Kumakura, *Polymer, 31*, 432 (1990).
[11] J. Nieuwenhuis, *Clin. Mater., 10*, 59 (1992).
[12] B. Buchholz, German Patent 443,542 A2 (1991).
[13] H. R. Kricheldorf and I. Kreiser-Saunders, *Makromol. Chem., 191*, 1057 (1990).
[14] Z. Jedlinski, W. Walach, P. Kurcok, and G. Adamus, *Ibid., 192*, 2051 (1991).
[15] J. Dahlmann, G. Rafler, K. Fechner, and B. Mehlis, *Br. Polym. J., 23*, 235 (1990).
[16] R. Dunsing and H. R. Kricheldorf, *Makromol. Chem., 187*, 1611 (1986).
[17] F. E. Kohn, F. G. Van Omnen, and J. Feijen, *Eur. Polym. J., 19*, 1081 (1983).
[18] H. K. Kricheldorf, M. Berl, and N. Scharnagl, *Macromolecules, 21*, 286 (1988).
[19] H. R. Kricheldorf and M. Sumbél, *Eur. Polym. J., 25*, 5851 (1989).
[20] P. Dubois, R. Jérôme, and P. Teyssié, *Makromol. Chem., Macromol. Symp., 42/43*, 103 (1991).
[21] P. Dubois, C. Jacobs, R. Jérôme, and P. Teyssié, *Macromolecules, 24*, 2266 (1991).
[22] L. Trofimoff, T. Aida, and S. Inoue, *Chem. Lett.*, p. 991 (1987).
[23] F. E. Kohn, J. W. A. Van de Berg, G. Van de Ridder, and J. Feijen, *J. Appl. Polym. Sci., 29*, 4265 (1984).
[24] N. Marcotte and M. F. A. Goosen, *J. Control. Rel., 9*, 75 (1989).
[25] K. Avgoustakis and J. R. Nixon, *Int. J. Pharm., 70*, 77 (1991).
[26] J. W. Leenslag and A. J. Pennings, *Makromol. Chem., 188*, 1809 (1987).
[27] P. B. Deasy, M. P. Finan, and M. J. Meegan, *J. Microencap., 16*, 369 (1989).
[28] Y. B. Lyudvig, B. G. Belen'Kaya, I. G. Barskaya, A. K. Khomyakov, and T. B. Bogomolova, *Acta Polym., 34*, 11 (1983).
[29] A. Duda and S. Penczek, *Macromolecules, 23*, 1636 (1990).
[30] H. R. Kricheldorf and A. Serra, *Polym. Bull., 14*, 497 (1985).
[31] F. Chabot, M. Vert, S. Chapelle, and P. Granger, *Polymer, 24*, 53 (1983).
[32] J. Leray, M. Vert, and D. Blanquaert, French Patent 76 28183 (1976).
[33] J. W. Leenslag, A. J. Pennings, R. R. M. Bos, F. R. Rozema, and G. Boering, *Biomaterials, 8*, 311 (1987).
[34] E. Lillie and R. C. Schultz, *Makromol. Chem., 176*, 1901 (1975).
[35] A. Schindler and D. Harper, *J. Polym. Sci., Polym. Lett. Ed., 14*, 729 (1976).

[36] M. Bero, J. Kasperczyk, and Z. J. Jedlinski, *Makromol. Chem., 191*, 2287 (1990).

[37] J. Mauduit, N. Bukh, and M. Vert, *J. Control. Rel., 23*, 221 (1993).

[38] M. Vert, in *Degradable Materials* (S. A. Baremberg, J. L. Brash, R. Narayan, and A. E. Redpath, Eds.), CRC Press, Boca Raton, Florida, 1990, p. 11.

[39] C. G. Pitt, F. I. Chasalow, Y. M. Hibionada, D. M. Klimas, and A. Schindler, *J. Appl. Polym. Sci., 26*, 3779 (1981).

[40] K. J. Zhu, R. W. Hendren, K. Jensen, and C. G. Pitt, *Macromolecules, 24*, 1736 (1991).

[41] S. M. Li, H. Garreau, and M. Vert, *J. Mater. Sci., Mater. Med., 1*, 123 (1990).

[42] S. Gogolevski, M. Javonovic, S. M. Perren, J. C. Dillon, and M. K. Hughes, *J. Biomed. Mater. Res., 27*, 1135 (1993).

[43] I. Grizzi, H. Garreau, S. M. Li, and M. Vert, *Biomaterials*, In Press.

[44] S. M. Li, H. Garreau, and M. Vert, *J. Mater. Sci., Mater. Med., 1*, 131 (1990).

[45] S. J. Holland, B. J. Tighe, and P. L. Gould, *J. Control. Rel., 4*, 155 (1986).

[46] Y. Doi, K. Kumagai, N. Tanahashi, and K. Mukai, in *Biodegradable Polymers and Plastics* (M. Vert, J. Feijen, A. C. Albertsson, G. Scott, and E. Chiellini, Eds.), Royal Society of Chemistry, Cambridge, 1992, p. 139.

[47] M. S. Reeve, S. P. McCarthy, M. J. Downey, and R. A. Gross, *Macromolecules, 27*, 825 (1994).

[48] M. Vert, A. Torres, S. M. Li, S. Roussos, and H. Garreau, in *Studies in Polymer Science, Vol. 12: Biodegradable Plastics and Polymers* (Y. Doi and K. Fukuda, Eds.), Elsevier, Amsterdam, 1994, p. 11.

SYNTHESIS OF POTENTIALLY BIODEGRADABLE POLYMERS

ZBIGNIEW JEDLINSKI and PIOTR KURCOK

Institute of Polymer Chemistry
Polish Academy of Sciences
41-800 Zabrze, Poland

ROBERT W. LENZ

Polymer Science and Engineering Department
University of Massachusetts
Amherst, Massachusetts 01003 USA

ABSTRACT

Homopolymers of β-butyrolactone, L-lactide, δ-valerolactone, and methyl methacrylate, as well as their block copolymers, were prepared with simple initiator systems based on alkali metal alkoxides and alkali metal supramolecular complexes. The goal was to synthesize polymeric materials with potential biodegradability.

INTRODUCTION

In past decades a great deal of research has been done on the synthesis of polymers and polymeric materials for long-term service. In particular, many kinds of thermally stable and chemically stable polymeric materials have been developed. Essential contributions to this field were made by Professor Herman Mark, with his remarkable ideas and research carried out by himself and his students, as well as by many other outstanding polymer scientists.

Many of the traditional applications of synthetic polymers are due to their chemical and biological inertness, as some of the earliest studies on the biodegradation of polymers were performed to prevent or to retard polymer degradation by microorganisms, such as bacteria and fungi. Today however, a new field of research has developed, which is concerned with the synthesis of biodegradable polymers. This field of research is growing rapidly because of the strong demand for such polymers as packing materials and, in particular, as biocompatible materials for specific medical applications, such as sutures, surgical implants, and formulations of drugs with controlled-release. Research on these materials continues at an extraordinary pace. One of the important current incentives for the study of biodegradable polymers is, of course, the potential for their easy disposal and biodegradation under natural environmental conditions.

Current attempts to develop biodegradable polymers have focused on the synthesis of new polymers and on the modifications of natural polymers. However, such materials are often too expensive for commercial applications, and, therefore, further studies are needed. One important commercial product is poly(β-hydroxybutyrate-β-hydroxyvalerate) (PHB/V) copolymer, which is produced via fermentation by Zeneca Bio Products in the UK and distributed under the tradename of Biopol.

Other synthetic polyesters obtained via the polymerization of lactones, lactides, cyclic carbonates, etc. are of interest for such application, and in our laboratory a new procedure has been elaborated for the synthesis of biodegradable "tailormade" polyesters with well-defined mechanical, physical, and chemical properties. In these syntheses anionic initiators are employed, capable of producing homopolymers, block polymers, or graft polymers via "living" polymerization processes. The interdisciplinary approach has been adopted for preparation of such initiators, which are based on the principles of supramolecular chemistry. The utility of such novel initiator systems, containing cation complexing agents as crown ethers, cryptands, and others, is discussed in this report.

SYNTHESIS OF HOMOPOLYMERS VIA RING-OPENING POLYMERIZATION WITH ALKALI METAL ALKOXIDES AS INITIATOR

Poly(β-Hydroxybutyrate) of Various Tacticities Synthesized with Alkali Metal Alkoxides/18-Crown-6 Complexes as Initiators

High molecular weight, stereoregular poly(β-butyrolactone)s have been synthesized with coordinative initiators [1, 2], but anionic polymerization reactions of racemic β-butyrolactone are also convenient methods in the synthesis of poly(β-hydroxybutyrate) with the desired tacticity. Furthermore, the results presented in this report indicate that the anionic polymerization of β-butyrolactone proceeds to almost complete conversion of monomer, both in bulk and in solvent, when alkali metal alkoxide/18-crown-6 complexes are used as initiators. End-group analysis by ^{1}H NMR revealed that, besides carboxylate active species, unsaturated and hydroxyl dead groups are present in the poly(β-butyrolactone) as prepared (Fig. 1) [3].

The tacticity of the poly(β-butyrolactone)s obtained in the presence of various alkoxides has been studied using ^{13}C-NMR spectroscopy (Fig. 2). The results obtained indicate that poly(β-butyrolactone) prepared in the presence either of potassium methoxide- or *tert*-butoxide/18-crown-6 complex possesses stereosequences distribution obeying Bernoullian statistics and are atactic. However, in polymeriza-

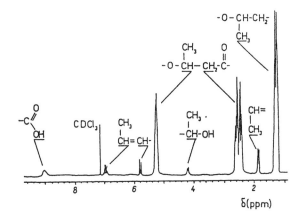

FIG. 1. ^{1}H-NMR of atactic poly(β-butyrolactone) prepared with potassium methoxide/18-crown-6 complex as the initiator.

tion with the optically active potassium (+)*sec*-butoxide/18-crown-6 complex, the product was predominantly syndiotactic poly(β-butyrolactone), in which with syndiotactic fraction was as high as 60%. The results of anionic polymerization of racemic β-butyrolactone initiated with various alkali metal alkoxide/18-crown-6 complexes are presented in Table 1. As shown by the data in Table 1, high molecular poly(β-butyrolactone) with a number-average molecular weight M_n, up to 40,000 and with various tacticities could be obtained via anionic polymerization of racemic β-butyrolactone with alkali metal alkoxide/18-crown-6 complexes as the initiator.

Synthesis of Poly(α-Hydroxy Acids) via Anionic Polymerization of L- and L,D-Lactide Initiated with Potassium Methoxide

The anionic polymerization of optically active L-lactide and racemic L,D-lactide in THF solution at room temperature proceeded fast and with good yields. The analysis of ^{1}H-NMR spectra of polylactides as obtained (Fig. 3) indicate that

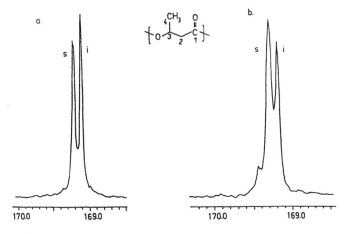

FIG. 2. ^{13}C-NMR (75 MHz) spectra (carbonyl carbon region) of (a) atactic poly(β-butyrolactone) and (b) predominantly syndiotactic poly(β-butyrolactone).

JEDLIŃSKI, KURCOK, AND LENZ

TABLE 1. Anionic Polymerization of Racemic β-Butyrolactone Initiated with Alkali Metal Alkoxide Complex with 18-Crown-6 at 20°C

No.	Potassium alkoxide	Solvent	$[M]_0$ in mol/dm^3	$[I]_0 \times 10^{-2}$ in mol/dm^3	Time in hours	Yield in %	M_n [a]	M_n (GPC)	M_w/M_n
1	MeOK	THF	2.30	6.80	3	98	2,900	3,100	1.08
2	MeOK	THF	3.62	1.02	150	96	30,500	24,000	1.07
3	MeOK	Benzene	1.89	7.40	4	99	2,200	4,100	1.09
4	MeOK	—	12.18	5.23	90	95	20,000	18,100	1.03
5	(+)sec-BuOK	—	12.18	29.93	6	98	3,500	3,600	1.10
6	(+)sec-BuOK	—	12.18	15.40	10	99	6,800	6,700	1.07
7	tert-BuOK	THF	2.5	3.07	10	98	7,000	6,900	1.05
8	tert-BuOK	THF	4.5	2.42	48	97	16,000	15,800	1.08
9	tert-BuOK	Benzene	3.42	1.73	48	98	17,000	16,900	1.07
10	tert-BuOK	—	12.18	4.70	85	99	22,500	21,300	1.08
11	tert-BuOK	—	12.18	2.30	170	95	45,000	38,000	1.08

[a]Number-average molecular weight calculated for 100% conversion.

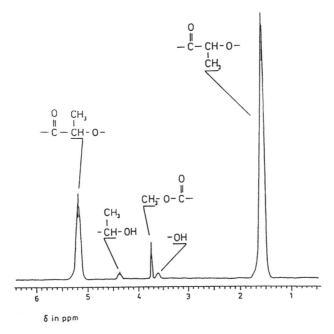

FIG. 3. ^{1}H-NMR spectrum of poly(L-lactide) prepared with potassium methoxide initiator.

the initiator was incorporated into the polymer chain in contrast to previous reports [4]. The results of anionic polymerization of lactides initiated with potassium methoxide are presented in Table 2.

The microstructure of the polymers obtained was studied using ^{13}C-NMR spectroscopy. The spectrum of poly(L-lactide) obtained (Fig. 4) in the region of methine and carbonyl carbons contains peaks corresponding to isotactic tetrads and hexads as well as signals corresponding to other stereosequences, although their intensities are small. Those results indicate that the polymers obtained possessed a high degree of tacticity, and revealed that the anionic polymerization of lactides

TABLE 2.   Anionic Polymerization of L- and L,D-Lactide with Potassium Methoxide in THF at 20°C

Lactide	$[M]_0$ in mol/dm^3	$[I]_0 \times 10^2$ in mol/dm^3	Time in minutes	Yield in %	M_{ncalc}[a]	M_n	M_w/M_n
L-	1.31	2.33	10	89	7,640	7,100	1.35
L-	1.66	1.63	15	90	13,300	13,500	1.30
L-	2.07	0.74	135	94	37,500	38,000	1.38
L,D-	0.98	1.86	15	90	6,960	6,500	1.37
L,D-	1.13	1.40	20	96	11,300	11,500	1.36

[a]Calculated molecular weight $M_{ncalc} = 144([M]_0/[I]_0) \times$ conversion.

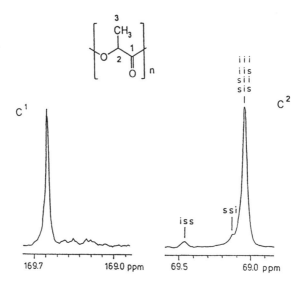

FIG. 4. ^{13}C-NMR (75 MHz) spectrum of poly(L-lactide) (carbonyl and methine region) prepared with potassium methoxide initiator; assignments for tetrads are indicated.

initiated with potassium methoxide led to well-defined polymers. Side reactions, such as racemization and transesterification, were negligible [5].

SYNTHESIS OF BLOCK POLYMERS VIA RING-OPENING POLYMERIZATION

δ-Valerolactone–L-Lactide Block Copolymers Obtained via Anionic Polymerization Initiated with Potassium Methoxide

The highly crystalline poly(α-hydroxyester)s obtained via lactide polymerization show poor water permeability and poor compatibility with soft tissue. This problem was overcome by block copolymerization of either glycolide or L-lactide with ε-caprolactone to yield polymers with good permeability and compatibility [6].

Previously, alkali metal alkoxides were found to act as initiators for the polymerization of L-lactide [5] and δ-valerolactone [7]. The polymerization of both monomers and the synthesis of their AB copolymers is demonstrated in this report. First, δ-valerolactone prepolymer was obtained by anionic polymerization initiated with potassium methoxide. This prepolymerization yielded "living" poly-δ-valerolactone with the expected molecular weight. When a THF solution of L-lactide was introduced into δ-valerolactone prepolymer solution, further reaction proceeded spontaneously, and after 15 minutes the L-lactide was almost entirely consumed, producing block copolymer as evidenced by GPC. The absence of peaks for possible cyclic oligomers in the GPC chromatogram of the reaction mixture suggests that no backbiting reaction took place in this process [8].

The ^{1}H-NMR spectrum of low molecular weight block copolymer obtained is presented in Fig. 5, and the results obtained for the block copolymerization reac-

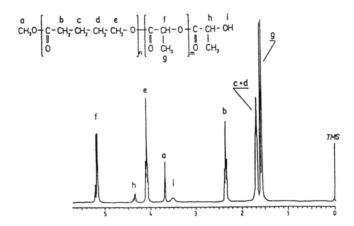

FIG. 5. ^{1}H-NMR (300 MHz) spectrum of poly(δ-valerolactone-*block*-L-lactide) prepared with potassium methoxide initiator (M_n = 3200, VPO; composition VL:LL = 55:45).

tions in THF at various ratios of δ-valerolactone prepolymer to L-lactide monomer are presented in Table 3.

The results in Table 3 indicate that poly-δ-valerolactone–poly-L-lactide block copolymers with the expected molecular weights were obtained with good yields, and the block copolymer compositions (i.e., content of δ-valerolactone and L-lactide segments) were close to the calculated values. Typical DSC traces for poly(δ-valerolactone-*block*-L-lactide) copolymers with different compositions are shown in Fig. 6. Two melting endotherms (Fig. 6, Curve b) are characteristic for block copolymers with a lactide content in the range between 43 and 62 mol%, which indicates that microphase separation of crystalline domains occurred in these poly-(δ-valerolactone-*block*-L-lactide)s.

TABLE 3. Preparation of Block Copolymers of δ-Valerolactone (VL) and L-Lactide (LL) with Potassium Methoxide in THF at 20°C[a]

Sample no.	Mole ratio of monomers in reaction mixture, VL:LL	Yield in %	M_n of prepolymer[c]	Polymer composition (mole ratio of units), VL:LL	M_{ncalc}[b]	M_{nGPC}[c]	M_w/M_n[c]
						of copolymer	
1	80:20	98	12,700	81:19	17,160	16,200	1.39
2	58:42	99	12,200	57:43	24,900	23,500	1.40
3	51:49	98	11,600	55:50	25,000	23,900	1.41
4	39:61	97	8,000	38:62	25,200	24,500	1.45
5	20:80	99	3,500	19:81	23,600	22,100	1.40

[a]The conversion of δ-valerolactone in the polymerization was higher than 99% after 5–10 minutes.
[b]Calculated for 100% conversion.
[c]Determined by GPC in THF.

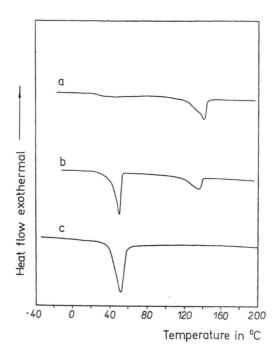

FIG. 6.   Typical DSC curves for original samples of block copolymers (a) VL:LL = 19:81, (b) VL:LL = 57.43, and (c) VL:LL = 81:19.

Synthesis of Poly(L-Lactide-*block*-Oxyethylene-*block*-L-Lactide) via Anionic Polymerization

The low hydrophilicity of poly-L-lactides decreases their compatibility with soft tissues and lowers their biodegradability [9]. This problem may be overcome by introducing hydrophilic segments into the polylactide chain, and in the present studies oxyethylene–L-lactide triblock ABA copolymers were synthesized. Sodium salts of poly(ethylene glycol)s were employed as macroinitiators for the anionic polymerization of L-lactide in THF at room temperature for this purpose.

It was found that the polymerization reaction proceeded rapidly, and after addition of the macroinitiator, the lactide monomer was almost entirely consumed in a short time. The block structure of the product obtained was confirmed by GPC and also by selective extraction experiments [10].

The results of ^{1}H-NMR measurements show that the polymerization of L-lactide initiated by the sodium salts of poly(ethylene glycol)s as macroinitiator yielded ABA-type block copolymers (Fig. 7).

The presence of a $-CH(CH_3)OH$ end group in the poly(L-lactide-*block*-oxyethylene-*block*-L-lactide) indicates that, similarly to the L-lactide polymerization initiated with potassium methoxide [5], the reaction proceeds via acyl–oxygen bond cleavage and that alcoholate anions were responsible for propagation. The results of the block polymerizations are presented in Table 4.

The results of water absorption experiments presented in Fig. 8 indicate clearly that the equilibrium water content in the poly(L-lactide-*block*-oxyethylene-*block*-L-

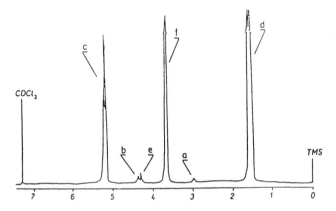

FIG. 7. ^{1}H-NMR (300 MHz) spectrum of poly(L-lactide-*block*-oxyethylene-*block*-L-lactide) prepared with sodium poly(ethylene glycol)ate initiator (M_n = 4800; composition EO:LA = 60:40).

TABLE 4. Anionic Polymerization of L-Lactide (LL) Initiated with Sodium Poly(Ethylene Glycol)ates (PEO) in THF at 20°C[a]

Sample no.	Mole ratio of EO to LL in the reaction mixture	Yield in %	Copolymer composition (mole ratio of units EO:LL)	Number-average molecular weight of block copolymer	
				M_{ncalc}[b]	M_n[c]
1	26:74	93	30:70	81,000	70,000
2	32:68	94	33:67	64,500	62,000
3	37:63	91	38:62	53,200	51,400
4	49:51	95	52:48	35,200	32,600
5	60:40	90	62:38	26,000	24,300

[a]M_n(VPO) of PEO used to obtain the macroinitiator was 8100. The polymerization time was equal to 5 minutes.

[b]Calculated for 100% conversion.

[c]Estimated from ^{1}H-NMR.

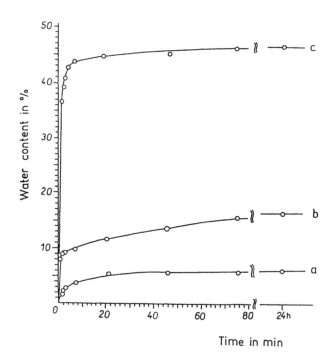

FIG. 8. Water absorption vs time curves for block polymers with compositions (a) EO:LA = 38:62, (b) EO:LA = 52:48, and (c) EO:LA = 70:30. For poly-L-lactide (M_n = 27,000), an equilibrium water content of 3% was determined.

lactide)s is considerably higher than that for poly(L-lactide), and the water content increased with increasing content of oxyethylene units in the block polymers obtained (Fig. 8).

The thermal properties of the block polymers obtained are presented in Table 5. The results in Table 5 indicate that some degree of crystallinity was present in the

TABLE 5. Thermal Properties of Poly(L-Lactide-*block*-Oxyethylene-*block*-L-Lactide)s[a]

Sample no.	Copolymer composition (mole ratio of units EO:LL)	M_n	$T_g{}^b$ in °C	$T_m{}^b$ in °C	$\Delta H_m{}^b$ in J/g	$\Delta H_{cryst}{}^b$ in J/g
1	30:70	70,000	20	153	38.3	30.1
2	33:67	62,000	29	151	32.8	25.3
3	38:62	52,400	18	145	31.2	23.0
4	52:48	32,600	9	137	18.6	18.6
5	62:38	24,300	−12	38	51.9	—

[a]Heating rate 10°C/min.
[b]T_g, T_m, ΔH_m, and ΔH_{cryst} were determined for original samples.

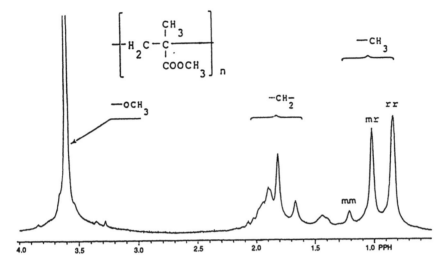

FIG. 9. ^{1}H-NMR spectrum of PMMA polymer prepared with supramolecular complex of sodium–potassium alloy as initiator in THF at 25°C.

block copolymers. The glass transition and the melting temperatures of the block copolymers depended on polymer composition and decreased when the oxyethylene unit content in the block copolymer was increased.

SYNTHESIS OF POLYMERS AND COPOLYMERS OF METHYL METHACRYLATE WITH ALKALI METAL SUPRAMOLECULAR COMPLEXES AS INITIATOR

Commercial poly(methyl methacrylate) is produced by radical polymerization, but it is also possible to synthesize stereoregular poly(methyl methacrylate)s. They are of better thermal stability than ordinary polymers obtained via radical

TABLE 6.   Homopolymerization of MMA in THF Initiated by Na$^-$/K$^+$, 18-Crown-6 Complex at 25°C. Initiator Concentration 0.03 mol/L

Sample no.	Yield, %	$10^{-3} M_{n,exp}$	M_w/M_n	Microstructure		
				i	h	s
1	100	18.0	1.3	8	37	55
2	100	36.0	1.6	9	36	55
3	100	37.7	2.5	10	37	53
4	100	33.6	2.2	9	39	52
5	100	37.7	2.5	10	37	53
6	100	66.1	2.6	9	38	53
7	98	106.0	3.2	8	38	54

TABLE 7.   Properties of Block Polymers of MMA and δVL in THF Initiated
by Na⁻/K⁺, 18-Crown-6 Complex at 25°C. Initiator Concentration 0.03 mol/L

Sample no.	Yield, %	$10^{-3} M_{n,exp}$	M_w/M_n	Polymer composition, % δVL[a]	T_g,[b] °C	T_m, °C	ΔH_m, J/g
1	85	33.5	2.4	14	115	51	3.8
2	85	34.0	2.3	48	–	62	43.6
3	87	36.5	2.1	68	–	54	47.8
4	91	24.3	2.5	83	–	63	61.6
5	93	30.2	2.4	91	–	60	66.1

[a]Estimated from ¹H-NMR.
[b]T_g for atactic PMMA = 105°C; T_g for syndiotactic PMMA = 125°C.

polymerization. In addition, coordinative and various anionic initiators [11–14]
have been used to prepare stereoregular polymers in order to improve the thermal
stability of methyl methacrylate polymers.

In our work, initiators based on solutions of modified alkali metal supramolecular complexes were employed. The first attempts to use alkali metal solutions as
initiators for methyl methacrylate polymerization were made by a French group
[15–17]. From the results of our previous work [18], and after some modifications

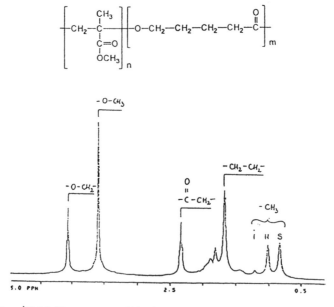

FIG. 10.   ¹H-NMR spectrum of block polymer prepared with supramolecular complex
of sodium-potassium alloy as initiator in THF at 25°C (M_n = 32,400; composition MMA:
VL = 54:46).

were carried out, very effective catalysts were prepared which work perfectly even at room temperature.

The ^{1}H-NMR (300 MHz) spectrum of PMMA prepared with these initiators is shown (Fig. 9), and the properties of the resulting polymers are summarized in Table 6. A high content of syndiotactic triads has been observed in polymers exhibiting molecular weights M_n up to 100,000.

The copolymerization of methyl methacrylate with δ valerolactone (δVL) initiated by "living" PMMA proceeds fast and produces block polymers exhibiting unimodal molecular weight distribution (Table 7). Results of selective extraction and ^{1}H-NMR spectra confirm the block structure of these polymers (Fig. 10), and the properties of the block polymers as obtained are summarized in the Table 7.

CONCLUSIONS

In the search for inexpensive methods of preparation of potentially biodegradable polymeric materials, simple initiator systems for cyclic monomers and methyl methacrylate, including alkali metal alkoxides with crown ethers as complexing agents and alkali metal supramolecular complexes, were investigated. The homo and block copolymers of β-butyrolactone, L-lactide, δ-valerolactone, and methyl methacrylate prepared with these initiators exhibited interesting physical properties and are expected to show biodegradability. Structure–property relationships for these materials were studied.

ACKNOWLEDGMENTS

The financial support from US-Polish M Skłodowska-Curie Joint Fund, Grant PAN/NSF-91-60, and from National State Committee for Scientific Research, Grant KBN 2 0867 9101, is acknowledged.

REFERENCES

[1]   R. A. Goss, Y. Zhang, G. Konrad, and R. W. Lenz, *Macromolecules, 21*, 2657 (1988).

[2]   J. E. Kemnitzer, S. P. McCarthy, and A. Gross, *Ibid., 26*, 1221 (1993).

[3]   P. Kurcok, Z. Jedliński, and M. Kowalczuk, *Ibid., 25*, 2017 (1992).

[4]   H. R. Kricheldorf and I. Kreiser-Saunders, *Makromol. Chem., 191*, 1057 (1990).

[5]   Z. Jedliński, W. Wałach, P. Kurcok, and G. Adamus, *Ibid., 192*, 2051 (1991).

[6]   C. X. Song and X. D. Feng, *Macromolecules, 17*, 2764 (1984).

[7]   K. Ito, M. Tomida, and Y. Yamashita, *Polym. Bull. (Berlin), 8*, 569 (1979).

[8]   P. Kurcok, J. Penczek, J. Franek, and Z. Jedliński, *Macromolecules, 25*, 2285 (1992).

[9]   R. A. Miller, J. M. Brady, and D. E. Cutright, *J. Biomed. Mater. Res., 11*, 711 (1977).

[10] Z. Jedliński, P. Kurcok, W. Wałach, H. Janeczek, and I. Radecka, *Makromol. Chem., 194*, 1681 (1993).

[11] S. K. Varshney, R. Jerome, P. Bayard, C. Jacobs, R. Fayt, and P. Teyssie, *Macromolecules, 25*, 4457 (1992).

[12] M. V. Beylen, S. Bywater, G. Smets, M. Szwarc, and D. J. Worsfold, *Adv. Polym. Sci., 86*, 87 (1988).

[13] B. Masar, P. Vlcek, J. Kriz, and J. Kovarova, *Macromol. Chem. Phys., 195*, 289 (1994).

[14] M. Kuroki, S. Nashimoto, T. Aida, and S. Inoue, *Macromolecules, 21*, 3114 (1988).

[15] S. Alev, F. Schué, and B. Kaempf, *J. Polym. Sci., Polym. Lett. Ed., 13*, 397 (1975).

[16] S. Alev, A. Collet, M. Viguier, and F. Schué, *J. Polym. Sci., Polym. Chem. Ed., 18*, 1155 (1980).

[17] M. Viguier, M. Abadie, B. Kaempf, and F. Schué, *Eur. Polym. J., 13*, 213 (1977).

[18] Z. Jedliński, A. Stolarzewicz, Z. Grobelny, and M. Szwarc, *J. Phys. Chem., 88*, 6094 (1984).

AEROBIC BIODEGRADATION OF SYNTHETIC AND NATURAL POLYMERIC MATERIALS: A COMPONENT OF INTEGRATED SOLID-WASTE MANAGEMENT

C. A. PETTIGREW, G. A. RECE, M. C. SMITH, and L. W. KING

The Procter & Gamble Company
Cincinnati, Ohio 45224

ABSTRACT

Solid waste collection, handling, and processing systems are changing worldwide as concerns increase regarding the environmental impact of solid waste. Recycling, waste to energy, and composting options are all expected to increase significantly as available landfill capacity decreases. As composting becomes an important method of managing solid waste, information about the biodegradability of polymeric materials and products is required in order to understand the fate of polymers in the environment. This paper describes the development of a tiered testing strategy for assessing the aerobic biodegradability and compostability of synthetic polymers. This testing strategy is demonstrated with polycaprolactone (PCL), a polymeric material that is generally regarded as biodegradable and compostable. PCL biodegradation was assessed in screening-level tests using the OECD 301B and ASTM D5338 test methods. In addition, ^{14}C-PCL was synthesized and used in confirmatory-level biodegradation tests that were conducted under realistic composting conditions. These studies show that screening-level biodegradation tests can provide knowledge of the inherent biodegradability of polymers, but the rate of biodegradation and the ultimate fate of polymeric materials in environmental matrices must be determined under realistic test conditions.

INTRODUCTION

Concerns about the management of municipal solid waste (MSW) continue to grow around the world. As waste management authorities are forced to deal with the problems associated with solid waste accumulating in landfills and with operation of mass-burn incineration systems, alternative solid waste collection, handling, and processing systems are being considered. An emerging principle that unifies these diverse options, called integrated waste management, requires that we process the various fractions of the solid waste streams according to the unique characteristics of the fractions. The application of this principle means that we will recycle those items that can be recovered for value, compost the organic biodegradable fraction that includes soiled paper, food, and yard waste, and incinerate the remaining organic fraction of the waste stream to recover energy. The remaining MSW, which will be small and largely inorganic (< 30–40% of the waste stream), can then be landfilled safely. In addition to recovering useful materials, the advantages of this approach include the use of smaller facilities for the same total waste flow, a better match between facility design and waste characteristics, less variability in the waste fed to the facilities, and the resultant residue is likely to be more stable than the original waste streams.

Composting will become increasingly important as a method of solid-waste management since this technique can handle up to 60–70% of the waste stream, converting the organic fraction into a useful product called compost [2]. Composting can be described as a method of solid-waste management whereby the organic component of the waste stream is biodegraded under controlled conditions. As composting becomes more prevalent, manufacturers will need to determine the compostability of their materials and products that will be disposed via solid waste. After consultation with technical experts and consideration of US Federal Trade Commission guidelines, we have defined a compostable material as:

> An organic material which undergoes physical, chemical, thermal, and/ or biological degradation in a municipal solid waste composting facility such that the material will break down into, or otherwise become part of, usable finished compost, and which ultimately will completely biodegrade in the environment in a manner similar to known compostable materials in municipal solid waste such as paper and yard waste.

One of the key elements of the compostable material definition requires that we assess the biodegradability of a test material. Biodegradation results when an organic material is used by microorganisms as a nutrient or when microorganisms carry out cometabolism. When microorganisms use the chemical as a carbon and energy source, they metabolize the chemical through biochemical reactions. Some of these reactions, which are catabolic, result in complete mineralization of the chemical to products such as CO_2 and CH_4 (under anaerobic conditions). Other reactions, which are anabolic, result in incorporation of the carbon, nitrogen, and other elements into new biomass.

During cometabolism, the chemical is not used as a nutrient and the microbial population responsible for biodegradation does not increase. Cometabolism typically occurs when a material is nonspecifically metabolized by enzymes meant for other compounds [4]. The result is usually only a partial breakdown of the material. On the basis of our review of the polymer biodegradation literature, we are not aware of any studies that clearly document the cometabolism of synthetic polymers.

In addition to being mineralized, another fate exists for carbon derived from chemicals that are biodegraded, particularly with polymers in compost and soil. This fate involves humification, in which metabolites as well as new and old biomass become incorporated into humic materials. This process is biologically and chemically catalyzed slowly over time and is a major sink for a significant portion of many organic compounds that enter compost and soil. Therefore, mineralization in soil, measured as the yield of CO_2 or CH_4 can be quite low [10].

Among the important criteria for compostability of new materials, biodegradation and disintegration need to be determined early in the evaluation process. We have developed a tiered testing approach toward the evaluation of new materials (Table 1). This approach allows us to focus the correct level of resources on the materials of greatest interest and potential. The objective of this paper is to describe screening-level and confirmatory-level test methods that can be used in the initial stages of the tiered testing approach and to highlight advantages and shortcomings with each method. These test methods are demonstrated with polycaprolactone (PCL), an aliphatic polyester that is generally regarded as biodegradable [3].

TABLE 1.   Tiered Testing Approach for Compostability Assessment

1. Screening
    A. Inherent Biodegradability
        Are the repeating subunits of a polymeric material inherently biodegradable?
        Is the polymer inherently biodegradable?
    B. Physical and Chemical Disintegration
        Does the material disintegrate under composting conditions?
        Do the chemical characteristics of the material change after exposure to a compost environment?
2. Confirmation
    A. Practical Biodegradability
        Does the material actually biodegrade in compost and soil under realistic conditions?
    B. Is the composted polymer/material safe?
3. Field Testing
    A. Fate and Effects Testing in MSW Compost Facilities
        Does the material/product show expected physical and chemical disintegration in situ over time?
    B. Fate and Effects Testing in Terrestrial Ecosystems
        Do persistent intermediates or residues accumulate over time?

MATERIALS AND METHODS

Sample Preparation

PCL resins were obtained from Union Carbide (Tone 787), and PCL films were 1–3 mil melt-blown films. PCL resins and films were pulverized with a Spex 6700 freezer mill (Spex Industries, Metuchen, NJ) using a stainless steel impactor, two steel end plugs, and a steel center cylinder. Film samples larger than the diameter of the vial were precut to approximately 0.5 cm squares. The grinding vial was filled with approximately 0.5 g of sample. The vial was placed in the cylindrical compartment within the mill and immersed in liquid nitrogen for approximately 1 minute, and the impactor was started. After 5 minutes of grinding the vial was removed from the liquid nitrogen and the sample particle size was determined. An additional 5 minutes of grinding was usually required. Sample surface area was determined using a Quantachrome Quantasorb Sorption System (Syosset, New York) that is based on gas absorption. The system was equipped with a Spectra-Physics SP4100 recorder (San Jose, California). Ultrahigh purity nitrogen was used for the adsorbate gas and ultrahigh purity krypton (0.078%) in helium as the carrier gas. Three contiguous replicates were tested for each sample. A Tekmar A-10 analytical mill (Cincinnati, Ohio) was also used to grind PCL film. The film was precut to approximately 0.5 cm squares. The mill lid and sample container, including sample, were cooled with a small volume of liquid nitrogen. After the liquid nitrogen evaporated, the mill was quickly assembled, pulsed for 10 seconds, unassembled, and the sample was brushed into the sample container. This procedure was repeated as necessary.

Radiolabeled-PCL was synthesized using [u-^{14}C]-ϵ-caprolactone prepared from ^{14}C-cyclohexanone (New England Nuclear) by a Baeyer–Villiger oxidation. The polymer was then prepared by bulk polymerization of a mixture of the labeled lactone and unlabeled caprolactone with a dialkylaluminum isopropoxide initiator in a manner similar to those described by Teyssié and coworkers [5]. The [u-^{14}C]-PCL was precipitated from methylene chloride in methanol and stored and used as a fluffy powder. Combustion analysis of the [u-^{14}C]-PCL was performed using a Packard 307 Sample Oxidizer (Meridin, Connecticut). Chromatographic analysis of the [u-^{14}C]-PCL was performed using a Water's Millennium Chromatography System (Marlborough, Massachusetts) equipped with a Model 410 differential refractometer detector, Shodex KF 804L columns, and a tetrahydrofuran mobile phase. The effluent from the columns was collected at 0.5 minute intervals and counted on a Beckman LS7800 (Fullerton, California) liquid scintillation counter to determine the distribution of radioactivity in the [u-^{14}C]-PCL sample.

Screening-Level Biodegradation Studies

The Sturm test, which measures CO_2 production as the biodegradation endpoint, is an aqueous aerated test that uses the supernatant from settled activated sludge as the source of microbial inoculum [9]. The modified Sturm test, designated 301B by the Organization for Economic Cooperation and Development [6], was used in this study. PCL samples were tested at a concentration of 110 mg/L in this test.

The compost-CO_2 test is an aerobic biodegradation test that uses stabilized mature compost derived from the organic fraction of municipal solid waste (MSW) as the microbial inoculum and the test matrix. This test simulates an intensive aerobic composting process and was done according to the procedure designated D5338 by the American Standards and Test Methods Society [1]. PCL was tested in triplicate at a concentration of 10% (w/w).

Confirmatory Biodegradation Studies

The extent of degradation of [u-^{14}C]-PCL was followed in a mature compost matrix that was equivalent to the compost used in the ASTM D5338 test method and was produced in a pilot-scale compost facility [8]. After adjustment of the moisture level to 50% (w/w), 100 g compost was added to 500 mL, three-neck, round bottom flasks. Each flask was fitted with a stirring shaft, a condenser, and an air sparger. The flasks were submerged in either a 45 or 55°C water bath, and experiments were conducted isothermally. Radiolabeled-PCL was added to test flasks at concentrations <1% (w/w) and tested in duplicate. A CO_2-free air supply was connected to each flask, and mineralized $^{14}CO_2$ was collected in a base trapping system. The flasks were connected to a CO_2 scrubbing train that consisted of two absorber bottles in series that were filled with 100 mL ethylene glycol monomethyl ether:ethanolamine (7:1). The CO_2 produced in each test flask reacted with the base in the absorber bottles, and the amount of CO_2 produced was determined by scintillation counting of the base traps. Periodically the absorber nearest the test flasks is removed for analysis. The remaining absorber bottle is moved one place closer to the test flasks and a new absorber bottle with fresh solution is added to the end of the trapping system. Scintillation counting was done using a Beckman LS7800 (Fullerton, California) operated in an external reference mode.

RESULTS AND DISCUSSION

Numerous methods have been developed to determine the biodegradability of natural and synthetic polymers. For example, Cain [3] reviewed the PCL biodegradation literature and noted that microbial growth, enrichment culture, and enzymatic depolymerization assays have been used to assess the biodegradability of this material. While all of these methods facilitate the elucidation of the mechanisms by which PCL is biodegraded, these methods generally produce results that are of limited value in assessing the environmental fate of a material and they do not provide direct evidence of complete biodegradation. In addition, the lack of consistent standard methods to assess polymer biodegradability has often resulted in confusion and misinterpretation among scientists and the lay public [7]. The tiered testing approach described in Table 1 was developed in order to assess the compostability and environmental safety of a large number of new synthetic polymers in an economically feasible manner. This tiered approach includes initial screening-level tests that determine the inherent biodegradability of a material followed by confirmatory laboratory and field testing where the practical biodegradability of a material is determined.

Screening-Level Studies

Knowledge of the inherent biodegradability of a polymer and its repeating subunits and degradation products is the first step of the tiered testing strategy described in Table 1. In some cases, such as the case for PCL, this information can be obtained by a literature review. However, the inherent biodegradability of new materials that have no prior testing history can be assessed by several screening-level tests. The modified Sturm test was originally developed in our laboratories in the early 1970s and is an example of a screening-level test [9]. Test compounds are placed in an inorganic medium and inoculated with the supernatant of homogenized activated sludge-mixed liquor. This test measures the mineralization of the test material, and the results are expressed as the percentage of theoretical CO_2 that could be produced if all of the carbon in the test material was mineralized to CO_2. The Sturm test was initially developed to assess the biodegradability of relatively low molecular weight materials that are discarded down the drain. Therefore, the Sturm test is ideal for testing the biodegradability of low molecular weight polymer repeat units or degradation products. However, due to the paucity of alternative methods, the Sturm test has often been used by default for testing high molecular weight polymer biodegradability. Because of the limitations associated with testing polymeric materials in the Sturm test, we investigated ways to improve polymeric material preparation and test reproducibility.

To study the effect of surface area on the biodegradability of PCL, we prepared a series of PCL samples with different surface areas and tested their biodegradability using the modified-Sturm test (OECD 301B). Increasing the surface area of PCL resulted in an increase in the rate and extent of CO_2 production for surface areas up to 0.122 m^2 per gram of sample (Table 2). Grinding PCL samples at liquid N_2 temperatures resulted in smaller particle sizes and larger surface areas without significant loss in molecular weight (Table 2). Despite optimizing the surface area of PCL, the maximum CO_2 yield obtained in this study was 63% TCO_2 after 180 days in the modified-Sturm test (Table 2). During this extended study the test chambers were reinoculated at periodic intervals in order to maintain an active microbial culture. While increasing the biomass level could facilitate the rate of biodegradation in the modified-Sturm test, the low biomass levels used in this test are required in order to obtain a good signal-to-noise ratio.

Since the conditions of the Sturm test were not originally optimized for poly-

TABLE 2. Effect of Surface Area on PCL Biodegradation in the Modified Sturm Test

Surface area (m^2/g)	MW (daltons)	Preparation technique	CO_2 production, % theoretical at 180 d
0.0035	109,500	Resin pellets	8.4
0.0112	99,400	Extruded film	36.6
0.0303	100,100	Film/analytical mill	37.4
0.0863	105,300	Pellets/cryogenic grinder	40.1
0.1216	104,300	Film/cryogenic grinder	62.6

meric materials, there are several limitations that must be considered. The length of polymer studies are often problematic since the microorganisms may die-off over time. The Sturm test was originally designed to run for 1 month, and tests requiring longer duration should be reinoculated. In addition, while an activated sludge inoculum is very diverse, it is primarily composed of bacterial populations, and the conditions of the Sturm test are not conducive to growth of filamentous microorganisms (e.g., actinomycetes and fungi). These groups of microorganisms are known to play key roles in the biodegradation of natural polymeric materials such as plant litter. Because of these limitations, alternative screening-level tests such as the ASTM 5338 Compost-CO_2 test were tried.

An alternative screening-level approach involves the use of a relatively high biomass level where the optimum CO_2 signal-to-noise ratio is maintained by increasing the test substance concentration. The biodegradability of a PCL sample ground with an analytical mill was tested in a compost matrix, containing a relatively broad diversity of microorganisms, using the ASTM D5338 test method. PCL degraded to an overall extent of 45% when considering all replicates (Fig. 1). However, one reactor acidified early in the test run and PCL biodegradation was halted. Biodegradation resumed after excess acid produced by the rapid hydrolysis of the PCL was degraded by the compost microorganisms. Neglecting the acidified reactor, the biodegradation after 45 days was 54% TCO_2. The ASTM D5338 method facilitates the exposure of the test material to a broad diversity of microorganisms in a high biomass system, but the high test substance concentration (10%, w/w) required to achieve good signal-to-noise ratios was problematic in the acidified reactor. Likewise, the artificially imposed temperature regime was noted to cause drastic shifts in microbial respiration (Fig. 1). While these results suggest that PCL is inherently biodegradable, they also point out the need for replicate analyses and demonstrate how different test conditions can yield different results in screening-level tests.

Confirmatory Studies

The second step of the tiered testing strategy outlined in Table 1 is the assessment of the practical biodegradability of a polymeric material. Biodegradation tests conducted at this level are designed to confirm the results that were obtained using

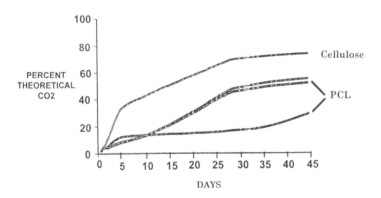

FIG. 1.   Biodegradation of PCL in the compost CO_2 test (ASTM D5338). The incubation temperature was controlled in order to simulate an intensive aerobic composting process (Days 0–1, 35°C; Days 1–5, 58°C; Days 5–28, 50°C; Days 28–45, 35°C).

screening-level test methods, and they attempt to simulate actual environmental systems. In the context of integrated waste management we chose to study the biodegradation of polymeric materials in an aerobic compost environment.

Biodegradation test conditions that mimic a compost environment involve high biomass levels in a solid matrix, low test substance concentrations, and require the use of sensitive analytical techniques. One common technique involves the use of isotopically labeled test materials. A uniformly ^{14}C-labeled PCL was prepared by "ring-opening" polymerization and was determined to have a specific activity 0.975 mCi/g. Chromatographic analysis of the [u-^{14}C]-PCL indicated that the M_n and M_w were 79,138 and 98,084, respectively, relative to polystyrene standards. The distribution of radioactivity in the various molecular weight fractions of the [u-^{14}C]-PCL was associated with a single peak, indicating radiochemical purity (Fig. 2). In addition, chromatographic analysis of the [u-^{14}C]-PCL indicated that only background levels of radioactivity were detected before and after the main peak.

The results of confirmatory-level biodegradation experiments using [u-^{14}C]-PCL in a mature compost matrix are shown in Figs. 3 and 4. These experiments were conducted under test conditions that mimic municipal solid waste compost conditions, and the results confirm the screening-level results that suggested PCL is inherently biodegradable. The combination of high biomass, relatively low concentration of test material, and elevated incubation temperatures facilitated the complete and relatively rapid biodegradation of PCL, indicating that PCL is practically biodegradable in a compost environment. Maintenance of isothermal conditions in these experiments facilitated kinetic analysis of the biodegradation data. The results of PCL biodegradation at 45 and 55°C are consistent with the general rule that higher temperatures that do not kill microorganisms or denature enzymes will result in higher metabolic activities. If the biodegradation of materials like PCL can be modeled by the Arrhenius equation, we may be able to predict the biodegradation of PCL and other similar polymers at ambient temperatures from studies run at elevated temperatures. However, in the case of PCL, the relative roles of biotic and

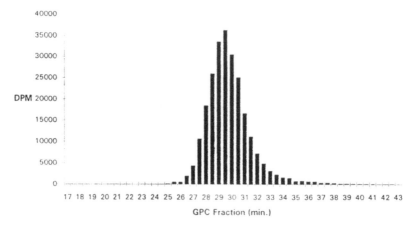

FIG. 2.   Distribution of radioactivity in fractions collected from a [u-^{14}C]-PCL sample separated by gel permeation chromatography.

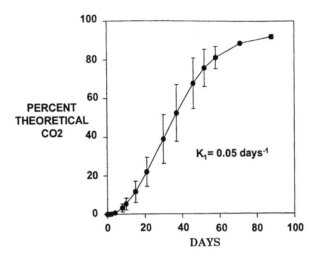

FIG. 3.   Biodegradation of PCL in 45°C compost $^{14}CO_2$ test. These data are the means and standard errors of the values obtained for two replicates. K_1 is the first-order rate constant (days^{-1}).

abiotic ester hydrolysis of PCL and the interactive effects that temperature may have with other factors such as T_m requires further testing.

Both the screening-level and confirmatory tests used in this study measured the mineralization of the test material to CO_2 as the primary biodegradation endpoint. While PCL was completely mineralized in the [u-^{14}C]-PCL compost test, many other slowly degrading synthetic polymers will require alternative biodegradation endpoints, such as disappearance of parent compound, appearance of interme-

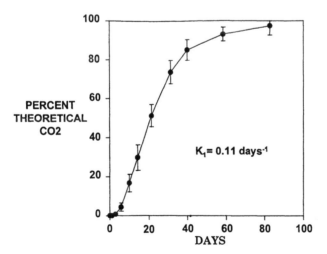

FIG. 4.   Biodegradation of PCL in 55°C compost $^{14}CO_2$ test. These data are the means and standard errors of the values obtained for two replicates. K_1 is the first-order rate constant (days^{-1}).

diates/metabolites, and incorporation into biomass or humic materials. Both the screening-level and confirmatory tests used in this study have the common limitation that they were in-vitro batch systems which ultimately encounter nutrient limitations and lack mechanisms for product removal. While closed batch systems facilitate mass balance determinations, the artificial conditions used in these tests do not mimic the open systems found in the environment and may ultimately limit the biodegradability of a test material. Alternative confirmatory-level biodegradation tests using mesocosm designs or field tests using in-situ conditions are required to achieve an open system.

CONCLUSIONS

As composting becomes an important method of managing solid waste, information about the biodegradability of polymeric materials and products is required in order to understand the fate of polymers in the environment. The implementation of a logical testing approach that includes consistent biodegradation test methods is critical to the development of compostable materials and testing claims of biodegradability. On the basis of initial experiments conducted with PCL, a synthetic polymer that is generally regarded as biodegradable, we conclude that screening-level tests provide useful information but that biodegradation test results can be highly dependent on the conditions of the test. Care must be taken to design screening-level tests to provide information on the inherent biodegradability of a polymer and not practical biodegradability. As noted in the compost-CO_2 test, attempts to generate both types of information tend to complicate the execution of the screening-level test and confuse the interpretation of results. The rate of biodegradation of complex high MW polymers can be very slow, and screening-level tests may provide misleading or false negative results. Confirmatory-level testing should be performed under realistic and relevant conditions in order to facilitate the extrapolation of results to the environment.

ACKNOWLEDGMENTS

The authors thank R. S. Honkonen, L. A. Schechtman, and J. R. Innis for the synthesis of [u-^{14}C]-PCL, and R. Herzog and S. W. Morrall for liquid chromatography analyses.

REFERENCES

[1]   ASTM, *Annual Book of ASTM Standards: Plastics (III),* Vol. 08.03, American Standards and Test Methods Society, Philadelphia, Pennsylvania, 1993, pp. 444–449.

[2]   Beyea, J., L. DeChant, B. Jones, and M. Conditt, "Composting Plus Recycling Equals 70 Percent Diversion," *BioCycle,* pp. 72–75 (May 1992).

[3]   Cain, R. B., "Microbial Degradation of Synthetic Polymers," in *Microbial Control of Pollution* (J. C. Fry, G. M. Gadd, R. A. Herbert, C. W. Jones,

and I. A. Watson-Criak, Eds.), Cambridge University Press, Cambridge, 1992, pp. 293–338.

[4]   Horvath, R. S., "Microbial Co-Metabolism and the Degradation of Organic Compounds in Nature," *Bacteriol. Rev., 36,* 146–155 (1972).

[5]   Jacobs, C., Ph. Dubois, R. Jerome, and Ph. Teyssié, "Macromolecular Engineering of Polylactones and Polylactides," *Macromolecules, 24,* 3027–3034 (1991).

[6]   OECD, *Guidelines for Testing Chemicals, Section III: Degradation and Accumulation, 301B,* Organization for Economic Cooperation and Development, Paris, 1981.

[7]   Palmisano, A. C., and C. A. Pettigrew, "Biodegradability of Plastics," *BioScience, 42,* 680–685 (1992).

[8]   Schwab, B. S., A. C. Palmisano, and D. J. Kain, "Characterization of Compost from a Pilot-Scale Composter Utilizing Simulated Solid Waste," *J. Waste Manage.,* In Press.

[9]   Sturm, R. N., "Biodegradability of Nonionic Surfactants: Screening Test for Predicting Rate and Ultimate Biodegradation," *J. Am. Oil Chem. Soc., 50,* 159–167 (1973).

[10] US Food and Drug Administration, *Environmental Assessment Technical Assistance Handbook PB87-1753345,* National Technical Information Service, Washington, D.C., 1987.

SYNTHESIS AND APPLICATIONS OF PHOTODEGRADABLE POLY(ETHYLENE TEREPHTHALATE)

J. E. GUILLET, H. X. HUBER, and J. A. SCOTT

Department of Chemistry
University of Toronto
Toronto, Ontario M5S 1A1, Canada

ABSTRACT

A novel synthetic route for the preparation of photodegradable poly(ethylene terephthalate) is described, along with procedures for the synthesis of ^{14}C-labeled plastic. Studies of the labeled plastic in burial tests in active soil show that significant levels of biodegradation are achieved over a 2 to 3 year period for both photodegraded and virgin resin. Computer modeling studies are used to assess the desirability and energy cost of various disposal strategies.

INTRODUCTION

Although the merits of "degradable plastics" as a means of solving some of the problems associated with the disposal of packaging materials in the solid waste stream remain to be demonstrated conclusively, their effective use in litter control is now well established. One of the most successful examples is the Hi-Cone beverage carrier, a rectangular sheet of ca. 15 mil polyethylene into which are punched six holes the size of conventional metal beverage containers. Weighing only a few grams, the manufacture of these packages makes few demands on our "nonrenewable" resources, and they replace plastic and paper carriers which are much more demanding in their resource and energy content. This minimalist design is a triumph of the concept of "under packaging."

For nearly a decade, a number of United States states, including California, have required this package to be photodegradable because it is highly litter-prone. It also was shown to endanger the lives of birds who often became entangled in the rings. After a number of years of development, an ethylene copolymer containing ca. 1% carbon monoxide was selected to manufacture the degradable variety of the plastic carrier. The resin is now available from two major plastics producers in the United States, and it is estimated that the total volume now exceeds 100 million pounds per annum for this one application. Furthermore, beach and other surveys of plastic litter show at least a tenfold reduction in the number of these carriers observed. In my own personal surveys of California beaches, the senior author can report that he has never seen one of these carriers since the legislation was passed. This confirms the predictions of the computer models we published more than 20 years ago [1]. The photochemistry of ethylene–CO copolymers was studied by Hartley and Guillet at the University of Toronto and published in the first issue of *Macromolecules* [2]. Its use in packaging had been patented by another former employee of Tennessee Eastman a few years earlier. Ethylene can be readily copolymerized with carbon monoxide by the high-pressure free-radical process. Compositions containing about 1% will photodegrade after about 3 weeks of exposure to outdoor sunlight and break up into small particles which are believed to biodegrade completely after photodegradation. At present no other photodegradable plastic can match the ethylene–CO copolymer on a cost/performance basis for this application.

The success of this product in reducing litter suggests the need for other types of photodegradable plastics for the manufacture of highly litter-prone packages, such as the Hi-Cone carriers and beverage containers. An obvious candidate would be photodegradable poly(ethylene terephthalate) (PET). PET can be made photodegradable by copolymerization with glycols or diacids containing ketone groups [3, 4]. Furthermore, masterbatches containing the ketone group can be prepared cheaply from recycled PET via a recently patented process [5] which is described in the Experimental Section. One possible application could be the Hi-Cone carrier, because a 3-mil film of PET would have the necessary physical strength to replace a 15-mil PE–CO, with considerable savings in material and energy costs.

Resource Implications

Both polyethylene and PET can be prepared from renewable resources such as corn or wheat. In fact, for a number of years, all of the world's polyethylene was synthesized from ethyl alcohol produced by fermentation of wheat or other cereals. The ethyl alcohol so produced was dehydrated to ethylene which was subsequently polymerized under high pressure to give polyethylene. Similarly, the chemical intermediates required to make PET could be synthesized from simple organic compounds available from natural products, but at much higher cost. However, the use of wheat or corn as a raw material for the production of plastics would require huge areas of land to be withdrawn from food production, a prospect that could not be morally defended in a world where population pressures have led to widespread starvation [1, 6].

Both PET and PE show accelerated rates of biodegradation after being subjected to extensive photodegradation prior to soil burial in an active environment.

EXPERIMENTAL

Preparation of Ecolyte PET Masterbatch

Photodegradable poly(ethylene terephthalate) (PET) was synthesized by the insertion of the di(ethylene glycol) ester of 4-ketopimelic acid (DGKP) into PET using a reactive extrusion process [5]. A Betol BTS 40 twin screw extruder, with contrarotating screws (40 mm diameter) and 21:1 L/D ratio, was used to conduct the insertion reaction of DGKP into molten PET.

The PET resin (Kodapak PET 9663 clear) was predried overnight and mixed with antimony trioxide (0.04 wt%) and dimethyl terephthalate (2.25 wt%). The mixture was loaded into an AccuRate feeder with the feed rate set at 1 kg/h. The temperature of the extruder barrel was kept in the 220–260°C range.

DGKP was fed into the PET melt stream through a vent port at a rate controlled to yield a 3-wt% addition level. The extruded filament was cooled in a water trough and pelletized.

The reaction product was thermally pressed into clear films and irradiated in a QUV Accelerated Weather Tester. The films failed at 95 hours whereas the control PET resin film remained intact.

Preparation of ^{14}C-Labeled Linear Polyethylene Masterbatch

The synthesis was carried out by the Ziegler process using a Parr pressure reaction apparatus, in hexane using titanium(IV) chloride and triethylaluminum as catalyst.

An ampule of ethylene-1 ^{14}C with an activity of 1 mCi was broken down by a standard vacuum technique into five bulbs containing 160 μCi each and one bulb with 200 μCi.

Ethylene ^{14}C was obtained from American Radiolabeled Chemicals, St. Louis, Missouri. Reagent-grade hexane (Caledon) was refluxed over metallic sodium under nitrogen. Titanium(IV) chloride (1.0 M solution in dichloromethane) and triethylaluminum (1.0 M solution in hexane) were obtained from Aldrich. Ethylene gas was from Matheson.

The pressure tank on the Parr pressure reactor was repeatedly flushed with ethylene gas to remove all traces of air. A double-valved bulb containing 160 μCi ethylene-1 ^{14}C was then connected between the ethylene cylinder and the Parr reactor, and the active ethylene was flushed into the pressure tank with ethylene gas. The bulb was then removed and the cylinder was connected directly to the pressure tank. The pressure was increased to 85 lb for the experiment.

The reaction vessel was loaded with 250 mL dry hexane. Titanium(IV) chloride solution (1.74 mL) and 2.9 mL of the triethylaluminum solution were added under nitrogen (0.2% of each catalysts, based on the hexane).

After reattaching the reaction vessel to the Parr pressure apparatus, the shaking mechanism was activated and the valve to the pressure tank was opened to keep the pressure in the reaction vessel at 40 to 45 lb. After 30 minutes the pressure in the tank dropped below 40 lb and the valve was opened completely. At this point the reaction had slowed considerably. After another 30 minutes no further pressure drop could be observed and the reaction was terminated. Any residual catalyst was destroyed with a mixture of propyl alcohol and methanol. The white, fibrous product was filtered and washed thoroughly with methanol and then dried in a vacuum oven at 60°C. The yield was 20.3 g.

The activity calculated from the amount of ethylene used (pressure drop from 85 to 23 lb = 62 lb or 73%) was 117 μCi for 20.3 g, or 5.75 μCi per gram. The activity found by scintillation counting was 5.92 μCi by combustion and 5.82 μCi by digestion. The product was stored under N_2 in brown glass bottles.

Experiments using this labeled PE in accelerated biodegradation tests using sewage sludge have been reported previously [7].

Synthesis of ^{14}C-Labeled PET

An ampule of ethylene glycol-1,2-^{14}C (Sigma) with an activity of 100 μCi was diluted to 10 mL with ethylene glycol (Fisher certified). This solution (1.0 mL) was mixed further with more ethylene glycol to a weight of 17.9 g (the amount calculated for this experiment) with a total activity of 10 μCi.

Step 1.   Transesterification.   The reaction was carried out in a 250-mL glass vessel equipped with a stainless steel stirrer, a high vacuum stirrer sleeve, a nitrogen inlet, and a take-off adapter. Dimethyl terephthalate (25.0 g, Aldrich 99+%), 8.0 mg magnesium acetate, and 13.2 mg Sb_2O_3 were introduced into the reaction vessel and heated while stirring at 60 rpm in an electric oven under nitrogen to 220°C. After generation and removal of methanol during 2.5 hours, 3.0 mg H_2PO_3 was added to deactivate the catalyst, and the temperature was raised to 250°C for 20 minutes. The resulting bis-hydroxyethyl terephthalate was allowed to cool to room temperature under nitrogen and was left overnight.

Step 2.   Polycondensation.   The condenser used in the removal of methanol in Step 1 was replaced by traps and the system was connected to a mechanical pump. After reducing the pressure to 0.35 mmHg, the reaction vessel was heated slowly to 280°C while stirring at 60 rpm. The distilling ethylene glycol was collected in an ice-water cooled trap. The second trap was cooled by liquid nitrogen. Initially the pressure increased to 0.56 mmHg, but it gradually decreased as the amount of free ethylene glycol decreased. During the reaction the stirrer power was gradually increased to maintain the speed. After 5 hours the reaction product was so viscous that stirring, even at full power, was no longer possible and heating was stopped. After cooling to room temperature, the product formed an extremely hard mass which stuck tightly to the stirrer and wall of the reaction vessel; it could not be removed by mechanical means. The product was dissolved in a 75:25 mixture of dichloromethane and trifluoroacetic acid, using a total of 250 mL of the solvent mixture. In order to retrieve the product (PET), thin films were cast from the solution and air dried, followed by vacuum drying at 60°C.

A total of 18.0 g PET was recovered, which represents a yield of 53%. The flexible films were stored in a brown glass bottle to protect them from exposure to light. The theoretical activity of 14.5 g, or 42.6% yield, is 4.26 μCi.

Samples for the biodegradation studies were prepared by solution blending 1.0 g of the ^{14}C-labeled PET (0.28 μCi) with 7.0 g of commercial PET (Hoechst) and 0.8 g PET containing 3.8% ketopimelate (Ecolyte PET masterbatch) in 100 mL of a 75:25 mixture of dichloromethane and trifluoroacetic acid. Thin films were prepared from this solution blend. They were air dried, followed by vacuum drying at 60°C. The average thickness of the films was 0.2 mm.

The films were separated into two lots: Lot #1 was stored in a brown glass bottle, and Lot #2 was exposed to the full output of a 275-W GE solar lamp. Lot #2 was placed directly under the source at a distance of 6 inches. After 30 hours the Lot #2 films were extremely brittle. Each lot was then milled in a Scienceware micromill using granulated sugar as a filler to increase the volume of material to the optimum volume required for grinding. The materials were milled for 15 minutes at liquid nitrogen temperature and washed into beakers with distilled water to remove the sugar.

After the sugar dissolved, the PET was filtered off, washed thoroughly, and dried in vacuum at 60°C. The dry powders were screened and passed entirely through a 50-mesh screen. Only traces (1-2%) were smaller than 100 mesh. Total amounts of sample obtained were called Sample A Ecolyte PET (unexposed), 1.27 g, and Sample B exposed for 30 hours at 6.0 inches, 1.50 g.

From each of the above samples, 0.5 g was thoroughly mixed with ca. 1.5 kg of soil mixture and placed in Tanks A and B. The activity of each 0.5-g sample was calculated to be 33 nCi. Actual determination showed 32 nCi, in excellent agreement.

A typical experimental setup is shown in Fig. 1. Three tanks were filled with ca. 1.5 kg of a mixture of unsterilized clay-loam garden soil, peat moss, beach sand, and fine gravel. Polymer samples were mixed into the top 0.5 cm of this mixture: Tank A, soil mixture plus 0.5 g of Sample A; Tank B, soil mixture plus 0.5 g of Sample B; Tank C, soil mixture only for determination of background radiation. Two seedlings each of *Fittonia verschaffeltii* (Hort.:Lem.) Coem., *Pellaea rotundifolia* (G. Forster) Hook., and *Oxalis* sp., were then planted in this mixture and 250 mL of 20% Hoagland's medium [8] was added.

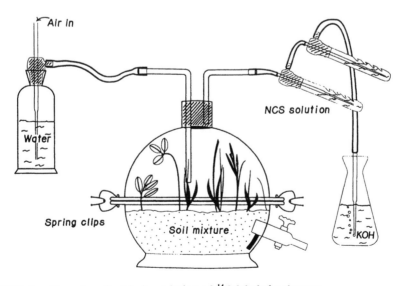

FIG. 1.   Terrarium for biodegradation of ^{14}C-labeled polymers.

The tanks were sealed and a constant flow of air was introduced at a rate of 10 mL/min. Exiting gas was bubbled through 1 M KOH to trap carbon dioxide (CO_2). At this rate of airflow, the free volume in each tank was replaced twice daily. Traps were changed every 2 weeks over the course of the experiment.

Short-Term Tests on Biodegradability

Tanks containing soil and labeled plastic samples were prepared. (For details of compositions, see Table 1.)

In the polyethylene experiments, the net accumulative radiation counts (in nCi) for Tanks A1, B1, and C1 are given in Fig. 2. It is clear that photodegradation causes a significant increase in the initial rate of biodegradation as measured by CO_2 evolution. Similar results were shown in the PET experiments (Fig. 3 and Table 2).

In each case the photodegraded plastic shows a more rapid rate of biodegradation than the undegraded material. There is no evidence that the presence of 16% starch in polyethylene causes any acceleration of biodegradation (Fig. 2C). In the case of PET, the unphotolyzed PET does not biodegrade in the first several months, but eventually degrades as rapidly as the photodegraded sample (Figs. 3 and 4).

TABLE 1.   Content of Tanks Used in Short-Term Tests

	Soil, dry wt, g	Plant, dry wt, g (final)	Plastic, wt, g	Form
			Experiment 1 – Polyethylene	
A1	1472.7	34.1	^{14}C-HDPE:LDPE:Ecolyte (20:75:5), 0.5000 g (= 330 nCi)	Powdered
B1	1512.5	21.1	^{14}C-HDPE:HDPE (20:80), 0.5000 g (= 375 nCi)	Powdered
C1	1399.3	30.8	^{14}C-HDPE:(HDPE + 16% starch) (20:80), 0.5000 g (= 380 nCi)	Film, 1 cm^2
D1	1506.0	26.4	No polymer added	Control
			Experiment 2 – Poly(Ethylene Terephthalate)	
A2	1642.4	18.8	^{14}C-PET, 0.5000 g (= 33 nCi)	Unexposed, powdered
B2	1559.7	23.6	^{14}C-PET, 0.5000 g (= 33 nCi)	Powdered, exposed 30 h at 6 in.
C2	1627.8	23.5	PET, 0.5000 g (= 0 nCi)	Control, unexposed, powdered

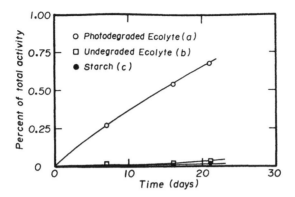

FIG. 2.   Biodegradation of polyethylene: accumulative activity of $^{14}CO_2$.

Long-Term Biodegradation Tests

Harvesting.   After about 2 years of plant growth, the tanks were opened, all plant material (including material adhering to the glass or plastic tank) and soil were separated and placed in preweighed, dry aluminum pans. Plant and soil samples were dried to constant weight in a 60°C oven.

Plant Extraction Procedure.   Dried plant samples were coarsely crushed and homogenized. Subsamples were taken and further homogenized under liquid nitrogen and redried. Five additional samples of 50 mg from each of the subsamples were combined (individually) with 15.0 mL toluene-based scintillation cocktail (2,5-diphenyloxazole (PPO, 4.0 g, New England Nuclear), 1,4-bis[2-(5-phenyloxazolyl)-benzene]) (POPOP, 0.5 g, Beckman), and scintillation grade toluene (1 L,

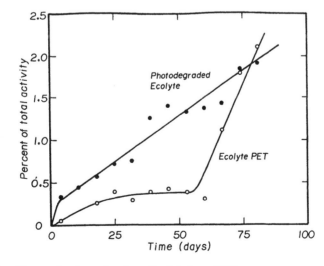

FIG. 3.   Net production of carbon dioxide from PET.

TABLE 2.   Biodegradation Study of ^{14}C Labeled PET

Days	Tank A (PET)		Tank B (Ecolyte PET) photodegraded	
	nCi[a]	%	nCi[a]	%
4	0.013	0.04	0.104	0.33
11	–	–	0.139	0.43
18	0.078	0.24	0.180	0.56
25	0.122	0.38	0.229	0.72
32	0.090	0.28	0.237	0.74
39	0.121	0.38	0.400	1.25
46	0.131	0.41	0.442	1.38
53	0.119	0.37	0.417	1.30
60	0.094	0.29	0.432	1.35
67	0.352	1.10	0.450	1.41
74	0.565	1.77	0.526	1.54

[a]After subtraction of counts from control Tank C.

Fisher) in glass scintillation vials. The samples were counted on a Beckman liquid scintillation counter.

Soil Extraction Procedure.   Soil (200 g) was added to either 50 mL MeOH or 50 mL 1:1 MeOH:distilled water in 125 mL Erlenmyer flasks. The flasks were sealed, placed on a rotary shaker set at 60 rpm, and extracted for 7 days. Samples were allowed to settle. The supernatant liquid was drained and centrifuged at 15 Krpm for 30 minutes. The supernatant extract was decanted and equally divided

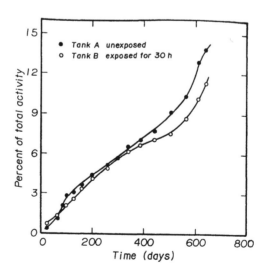

FIG. 4.   Biodegradation of PET.

into two glass scintillation vials (10.0 mL to each). The toluene-based scintillation cocktail described above was added to a total volume of 15 mL. The samples were counted on a Beckman liquid scintillation counter.

The long-term release of $^{14}CO_2$ is shown in Fig. 4. Total release from both photodegraded and undegraded PET is about 15% over 2 years, which compares favorably to many natural polymers. Furthermore, the rate of $^{14}CO_2$ evolution appears to accelerate as the degradation proceeds, which might be expected to occur as the polymer fragments break down in molecular weight.

Preliminary results on plant and soil assays are shown in Table 3. In the case of PET, plant assays indicate that the plants contain nearly as much ^{14}C as the CO_2 released. A similar result was observed in earlier studies with ^{14}C-labeled Ecolyte PS [9]. Taken together with the CO_2 results, this suggests that PET is more than 30% biodegraded after burial for 2 years in an active soil environment, a result comparable to those reported for straw by Jansson [10].

No appreciable amounts of radiolabeled residues were detected from either 50/50 water/methanol or methanol extraction of PET containing soils, suggesting that most low-molecular weight compounds produced are biodegraded more rapidly than their precursors. However, some methanol-soluble products were extracted from the linear polyethylene tests (Table 3). The composition of these residues will be investigated further.

Environmental Considerations

The long-term burial tests described herein strongly suggest that PET is inherently a truly biodegradable plastic. Furthermore, its return to the environmental carbon cycle when it becomes litter can be accelerated by minor modifications in its chemical structure to include small amounts of ketone carbonyl groups. This procedure can be carried out either in its original synthesis or in a subsequent reactive extrusion to form a masterbatch which can be pellet blended before extrusion to film, or blow molded to provide bottles or other containers.

Aluminum cans have been considered to be environmentally friendly for beverage containers because they are easy to recycle. But the same is true of PET, and

TABLE 3. Results of Long-Term Burial Tests (2 years)

| Composition | Morphology | Localization of ^{14}C (%) | | | |
		Soil[a]	Plant	Air (CO_2)	Total %
PET, unexposed	Powder	0	14	15	30
PET, exposed 30 h at 19 cm	Powder	0		14	14
HDPE:LDPE:Ecolyte					
(20:75:5)	Powder	5			
^{14}C-HDPE:HDPE (20:80)	Powder	10			
HDPE:HDPE containing					
16% starch (20:80)	Film, 1 cm^2	0			

[a]Based on methanol extraction at 25°C.

TABLE 4.   Energy Costs of Beverage Containers[a]

Container	MJ/kg	kWh/container
Aluminum can	305	2.4
PET bottle	107	0.84

[a]From I. Boustead and G. F. Hancock, *Energy and Packaging*, Ellis Horwood Publishers, Chichester, 1981.

its energy cost is much lower. Data on the relative energy costs of a typical aluminum can and a PET 450 mL bottle calculated from the data of Boustead and Hancock [11] are shown in Table 4. The aluminum can requires three times as much energy to manufacture, and if not recycled a similar waste of energy. Most new electrical plants in the United States use fossil fuels (gas, coal, oil) to generate the power for aluminum production, so that in this case the use of aluminum cans causes a much greater drain on nonrenewable resources than do plastic containers. Furthermore, an aluminum can is virtually indestructible in the environment and represents a form of "permanent litter." In view of the billions of such containers produced annually in the developed countries and the relatively low rate of recycling (< 40%), this is a consideration that urgently needs attention.

ACKNOWLEDGMENTS

The authors gratefully acknowledge the financial support of the Natural Sciences and Engineering Research Council of Canada, the Ontario Centre for Materials Research, and the Charles A. Lindbergh Fund, Inc. We thank Mr. John Braun of Ecolyte Atlantic Inc. of Baltimore, Maryland, for helpful discussions.

REFERENCES

[1]   J. E. Guillet, *Plast. Eng.*, pp. 47–56 (August 1974).
[2]   G. H. Hartley and J. E. Guillet, *Macromolecules, 1*, 165 (1968).
[3]   L. Alexandru and J. E. Guillet, *J. Polym. Sci., Polym. Chem. Ed., 14*, 2791 (1976).
[4]   J. E. Guillet and E. Dan, US Patent 3,878,169.
[5]   J. E. Guillet, I. Treurnicht, and R. S. Li, US Patent 4,883,857.
[6]   J. Guillet, *Polym. Mater. Sci. Eng., 63*, 630 (1990).
[7]   J. E. Guillet, H. X. Huber, and J. Scott, in *Biodegradable Polymers and Plastics* (M. Vert, J. Feijen, A. Albertsson, G. Scott, and E. Chiellini, Eds.), The Royal Society of Chemistry, Cambridge, 1992, pp. 55–70.
[8]   D. R. Hoagland and D. I. Aron, *Calif. Agric. Exp. Circ. 347* (1950).

[9]   J. E. Guillet, M. Heskins, and L. R. Spencer, *Polym. Mater. Sci. Eng., 58,* 80 (1988).

[10] S. L. Jansson, "Nitrogen Transformation in Soil Organic Matter," in *The Use of Isotopes in Soil Organic Matter Studies,* Report of the FAO/IAEA Technical Meeting, September 9–14, 1963, Pergamon Press, Oxford.

[11] I. Boustead and G. F. Hancock, *Energy and Packaging,* Ellis Horwood Publishers, Chichester, 1981.

PROPOSED MECHANISM FOR MICROBIAL DEGRADATION OF POLYACRYLATE

FUSAKO KAWAI

Department of Biology
Kobe University of Commerce
Gakuen-Nishimachi, Nishi-ku, Kobe 651-21, Japan

ABSTRACT

A possible aerobic degradation mechanism for polyacrylate (PA) was examined with acrylic oligomer-utilizing bacteria (*Microbacterium* sp., *Xanthomonas maltophilia*, and *Acinetobacter* sp.), using a model compound (1,3,5-pentane tricarboxylic acid, PTCA). Acyl-coenzyme A synthetase activities were detected with dialyzed cell-free extracts of PTCA-utilizing bacteria toward PTCA, PA 500, and PA 1000. This result suggested that PA is activated by coenzyme A and metabolized via PA-coenzyme A. Metabolic products formed from PTCA were detected in culture filtrates and reaction mixtures of washed cells. Fraction A was detected as a main metabolite by high-performance liquid chromatography. A small amount of fraction B was concomitant with fraction A. Also, another fraction, C, was detected. These intermediate metabolites were characterized by LC-MS as 1,3,5-(1- or 2-pentene)tricarboxylic acid for fractions A and B and as 1,3,5-(2-oxopentane)tricarboxylic acid for fraction C. Fraction A was metabolized far faster than fraction B. Fraction B was thought to be an artifact formed from fraction A under alkaline conditions. Thus PTCA and also PA seemed to be metabolized by the mechanism similar to β-oxidation of fatty acids. The degradation of PTCA by washed cells was slower than that by growing cells and was inhibited by 5 mM NaN_3. This suggests that the metabolism is linked to a respiratory chain of bacteria.

THE METABOLISM OF PTCA AS A MODEL FOR ACRYLIC OLIGOMERS

PTCA-utilizing bacteria (*Microbacterium* sp., *Xanthomonas maltophilia*, and *Acinetobacter* sp.) [1] grew on 2-methylglutarate (acrylic dimer) and PTCA, but appreciable amounts of metabolic products could not be detected in a culture supernatant or reaction mixture of intact cells with 2-methylglutarate. As accumulation of metabolic products from PTCA was found on HPLC, this compound was used as a model. No detectable amounts of significant artifacts were detected with an autoclaved PTCA medium. Then, culture supernatants of PTCA-utilizing bacteria were analyzed by HPLC. The main product was fraction A with a small amount of fraction B formed, as shown in Fig. 1 [2]. Although a large amount of fraction A was formed in 2–3 days, it was quickly further metabolized and disappeared from the culture supernatant after 4–5 days. Fraction B was only slowly metabolized and remained in the culture supernatant for an extended time (5–7 days). The same elution profiles as those with the culture supernatants of growing cells were obtained with reaction mixtures of washed cells, although the substrate was metabolized more slowly by washed cells than by growing cells. Substantially no difference in

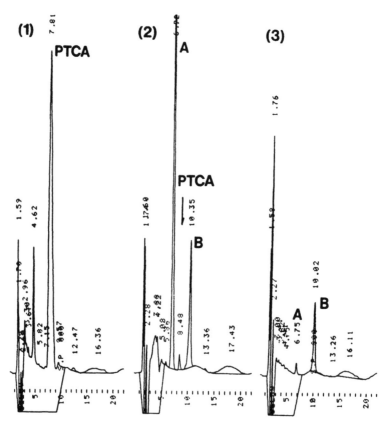

FIG. 1.   Time course for the consumption of PTCA and the amounts of Fractions A and B formed by growing cells of *Microbacterium* sp. (1) Autoclaved medium; (2) 3-day culture supernatant; (3) 7-day culture supernatant. Reprinted from Ref. 2.

the formation of metabolic products by the three bacteria was found with either culture supernatants or reaction mixtures. Biodegradation rates and accumulation of intermediate metabolites were affected by aerobic conditions. With increased shaking (120 rpm), approximately 100% of PTCA was consumed in 2 days, but small amounts of metabolites accumulated in the culture supernatant. With moderate shaking (100 rpm), considerable amounts of metabolites accumulated in 2–3 days in the culture supernatant.

Culture supernatants were analyzed by LC-MS on Waters or Cosmosil $5C_8$. By positive and negative ion detection, the molecular weights of fractions A and B were suggested to be the same, 202, which is 2 less than the molecular weight of the substrate, 204. Fractions A and B were eluted later than the substrate, suggesting that these compounds are less hydrophilic. Therefore, fractions A and B were consistent with 1,3,5-(1- or 2-pentene)tricarboxylic acid. Fraction B was found to be an artifact formed from fraction A at an alkaline site. Also, another peak was eluted on Cosmosil 5CN-R. The molecular weight of fraction C was considered to be 217 by negative ion detection. Considering the β-oxidation pathway, 1,3,5-(2-oxopentane)tricarboxylic acid is consistent with fraction C.

From these results, the metabolic pathway for PTCA is proposed, as shown in Fig. 2. Considering a β-oxidation mechanism is working for PTCA, fraction A as the main metabolite might correspond to 1,3,5-(1-pentene)tricarboxylic acid. Two possibilities still remain for the metabolic pathway of PTCA: 1) decarboxylation of a depolymerized compound by one acrylic unit followed by the ordinary β-oxidation process to liberate acetic acid and 2) repeated liberation of malonic acid from the depolymerized compound. Malonic acid is known as a potent inhibitor for succinate dehydrogenase in eucaryotic cells. As three kinds of PTCA-utilizing bacteria utilized malonic acid as a growth substrate, malonic acid is nontoxic to these procaryotic cells.

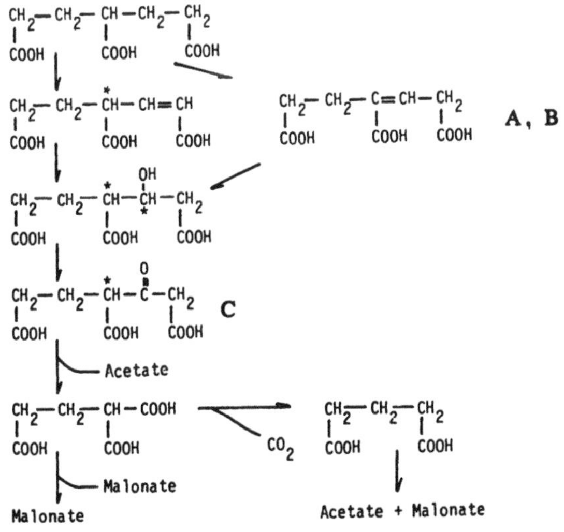

FIG. 2.   The proposed metabolic pathway for PTCA. Reprinted from Ref. 2.

PROPOSED METABOLIC PATHWAY FOR PA

As PTCA and PA 500 and PA 1000 can be activated by acyl-coenzyme A synthetase which is a key enzyme for starting the β-oxidation of saturated fatty acids and organic acids [1]. As described above, the characterization of metabolic products also suggested that enzymes similar to those for β-oxidation are involved in the metabolism of PTCA. PA 1000-3000 was metabolized by washed cells of PTCA-utilizing bacteria. As the oxidation site is known to be located on bacterial membranes, a polymer has to contact the metabolizing enzymes on a cytoplasmic membrane. Actually the degradation of PTCA by washed cells was completely inhibited by 5 mM NaN_3, an inhibitor for a respiratory chain, suggesting that the metabolism of PTCA proceeds on the cytoplasmic membrane. When considering a polymer, a short length of the chain might penetrate through the outer cell membrane or cell wall, although long chain polycarboxylic acids could hardly reach the cytoplasmic membrane. Therefore, it is quite reasonable to think that the metabolism of PA is exogenous and not randomly endogenous.

CONCLUSION

1. PTCA, PA 500, and PA 1000 were activated by coenzyme A.
2. The intermediate metabolites from PTCA, a model for the metabolism of PA, were identified as 1,3,5-(1- or 2-pentene)tricarboxylic acid and 1,3,5-(2-oxopentane)tricarboxylic acid.
3. From these results, PA seemed to be metabolized by the mechanism similar to β-oxidation of fatty acids.
4. The metabolism seemed to be linked with the respiratory chain of bacteria.
5. The degradation mechanism is possibly exogenous.

REFERENCES

[1] F. Kawai, *Appl. Microbiol. Biotechnol., 39,* 382 (1993).
[2] F. Kawai, K. Igarashi, F. Kasuya, and M. Fukui, *J. Environ. Polym. Degrad., 2,* 59 (1994).

COMPOSTABLE FILMS OF MATER-BI Z GRADES

C. BASTIOLI, F. DEGLI INNOCENTI, I. GUANELLA, and G. ROMANO

Novamont
Via Fauser, 8-28100 Novara, Italy

ABSTRACT

This paper focuses on the composting behavior of Mater-Bi Z grades for filming and on their in-use behavior, mainly as composting bags for the collection of organic waste. The compostability of Mater-Bi Z grades was demonstrated by means of a three-step test by following the approach of different international committees working on this matter comprising lab tests (biodegradability tests such as ASTM D 5338, ASTM 5901, and terrestrial toxicity tests) and full-scale composting tests. In addition to their compostability, the in-use behavior was estimated by tests of separate collection of organic wastes in different municipalities (Furstenfeldbruck/Bavaria, Korneuburg/Austria) comparing paper bags with Mater-Bi bags.

INTRODUCTION

Novamont produces three classes of biodegradable materials, A, Z, and V, under the Mater-Bi trade mark. They all contain thermoplastic starch, characterized by different synthetic components:

- A class products: biodegradable, not compostable because the degradation time is inconsistent with a composting cycle. They are mostly used for rigid-injected or blow-molded parts.
- Z class products: biodegradable and compostable. They contain aliphatic polyesters and are particularly suitable for the production of films, layers, and injected or blow-molded parts.
- V class products: biodegradable, compostable, and soluble. They are mostly used as a substitute for expanded polystyrene.

247

In all cases the transformation technologies for traditional plastics can be applied with minor modifications.

This paper focuses on the compostability of Mater-Bi Z grades for filming and on their in-use behavior, mainly as composting bags for the collection of organic wastes.

EXPERIMENTAL

All the data of this paper relate to Mater-Bi grades of the Z class: ZF02U, ZF03U, and ZI01U. They contain around 50% polycaprolactone plus starch and other natural plasticizers and compatibilizers.

Pure cellulose and different types of paper have been used as reference materials.

The compostability of these materials was demonstrated by means of a three-step test according to the approach followed by different international committees (ISR/ASTM, CEN, ORCA) working on this matter.

1. Intrinsic biodegradability of the material by standard respirometric tests simulating composting conditions (ASTM D5338) and watery environments such as the urban depurator (OECD modification of Sturm Test) [1–4].
2. Terrestrial Toxicity Tests (see germination [5], terrestrial plant growth [6], and worm acute toxicity [7]) and physical-chemical characterization of the compost obtained by Point 1, and prepared according to an ASTM procedure "for preparing residual solids. . . . "
3. Final cross-verification through full-scale composting trials. Full-scale composting plants covering different technologies (static windrows, turned windrows, rotary fermenting reactors) and different locations (Germany, Italy, Japan) have been considered.

Tensile and tear properties of these products are in line with those of LDPE, and the water permeability is one order of magnitude higher than that of LDPE.

Experimental projects of separate collections of organic wastes were performed in different municipalities (Furstenfeldbruck/Bavaria [8]; Korneuburg/Austria [9]; Padova, Brescia, Trento, etc./Italy) comparing paper bags with Mater-Bi bags to estimate, in addition to their compostability, their in-use behavior.

RESULTS

Compostability Test

Intrinsic Biodegradability

Modified Sturm Test. The test procedure is close to ASTM D5209-91 and to the test of the Italian decree DM 7/12/90. The test forecasts, in parallel, the analysis of well known reference substances (sodium acetate, Lissone paper for food contact) [1, 2]. All Z grades show a biodegradation index (CO_2/ThCO_2) at 56 days quite similar to that of the paper for food contact, according to what is required by the Italian ministerial decree DM 7/12/90 to define an unsoluble material as biodegradable.

Controlled Composting Test (ASTM D5338). The test simulates a composting environment and compares the biodegradation behavior of the test material with that of pure cellulose, both at concentrations of 10% w/w. In 45 days of testing the biodegradation indexes for ZF02U, ZF03U, and ZI01U were 74, 78, and 100%, respectively, against values around 85% obtained for pure cellulose [10, 11].

Terrestrial Toxicity Tests

The tests of seed germination, terrestrial plant growth, and worm acute toxicity, together with the physical-chemical characterization, performed by OWS [12] on the compost obtained by ASTM test D5338, gave results comparable with both the reference composts from the degradation of cellulose and for the control compost.

Full-Scale Composting Tests

The tests performed on different plants in Europe and Japan showed a complete disappearance of the material and an excellent compost quality. The main composting plants which tested Mater-Bi Z products were Scuola Agraria-Parco di Monza/Italy (static windrows) [13], Scuola Agraria-San Michele all'Adige/Italy (turned windrows), and Nogi-Cho Recycling Center/Japan (rotary fermenting reactor) [14].

As an example, Scuola Agraria of Parco di Monza composted pen caps of ZI01U in a static windrow, threaded with a plastic line, and entangled in a plastic tubular net. At the end of the cycle, the nets were empty and only residual lumps, stuck on the net with the typical appearance of compost, were recoverable. Other solid natural products in the heap, such as small pieces of wood, were yet not degraded, with only the surface being slightly attacked.

In-Use Behavior of ZF03U Composting Bags

The in-use behavior of 6-L ZF03U bags was evaluated by tests of separate collections of the organic wastes organized by different municipalities.

As an example, in Furstenfeldbruck Municipality/Bavaria, 55.7% of the citizens involved in the experiment were in favor of Mater-Bi bags whereas 28.6% were in favor of paper bags. In Korneuburg Municipality/Bavaria, 87% of the citizens preferred Mater-Bi bags. The compost quality was found to be in line with the national specifications for compost.

CONCLUSIONS

The results of full-scale composting tests and of standard lab tests performed by OWS on the biodegradation behavior of Mater-Bi Z grades and on the terrestrial toxicity of the resulting compost prove the compostability of these materials. Moreover from a functionality point of view, the experiments with separate collections organized in different European countries together with the physical-chemical characteristics of the film demonstrate the good in-use performances of Mater-Bi bags.

REFERENCES

[1]   G. P. Molinari, *Saggio di biodegradabilita' aerobica secondo D.M. del 07/12/1990 di ZI01U,* December 13, 1993.

[2]   G. P. Molinari and C. Freschi, *Saggio di biodegradabilita' aerobica secondo D.M. del 07/12/1990 di ZF03U,* March 1994.

[3]   G. Freschi, C. Bastioli, F. Degli Innocenti, G. Gilli, and G. P. Molinari, *Congress of Società Italiana Chimica Agraria,* 1994, In Press.

[4]   EEC Directive 79/831.

[5]   Zucconi et al., *BioCycle, 22,* 27–29 (1981).

[6]   OECD 207.

[7]   OECD 208.

[8]   *Landratsamt of Fürstenfeldbruck/Bavaria – Pilot Projekt – Mater-Bi Säcke,* Report, June 1993.

[9]   W. Hauer, *Vorarbeten für die Markteinführung von Säcken aus Mater-Bi zur Samulung von biogenen Abfällen,* Kornenburg, September 1993.

[10] J. Boelens, *Aerobic Biodegradation under Controlled Composting Conditions of Test Substance ZI01U,* OWS, September 30, 1992.

[11] J. Boelens, *Aerobic Biodegradation under Controlled Composting Conditions of Test Substance ZF03U,* OWS, July 31, 1992.

[12] B. D. Wilde and I. Boelens, *Compost Quality Tests of Test Substance ZF03U Compost,* OWS, November 6, 1992.

[13] E. Favoino and M. Centemero, *Compostability Testing of Injection Molded Items in Mater-Bi: General Information,* Scuola Agraria of Parco di Monza, May 24, 1993.

[14] Y. Yoshida and T. Vemura, in *3rd International Scientific Workshop on Biodegradable Plastics and Polymers,* Osaka, November 9-11, 1993, p. 72.

BIODEGRADATION OF POLY-D,L-LACTIC ACID POLYURETHANES

S. OWEN, M. MASAOKA, R. KAWAMURA, and N. SAKOTA

Central Laboratory
Rengo Co Ltd.
Ohhiraki 4-1-186, Osaka, Japan

ABSTRACT

The relationship between the biodegradability of poly-D,L-lactic acid (PLA) polyurethane compounds and their polymer composition was investigated. The biodegradability (weight loss) by fungi of these polyurethanes increased when: 1) the polymethylene polyphenyl polyisocyanate content of the polyurethane was reduced; 2) the molecular weight of the polyethylene glycol moiety in the polyol was increased; 3) the lactic acid content of the polyol was decreased. Proton NMR analysis of polyurethanes before and after biodegradation showed that the weight loss of PLA polyurethanes is mainly due to biodegradation of the polyol segment within the polyurethane. Measurement of oxygen consumption during cultivation of fungi indicated that only the polyol of PLA polyurethanes can be biodegraded to CO_2. However, a strain of microorganism capable of biodegrading a urethane compound (bisethylurethane of tolylene-2,4-diisocyanate) was isolated, and it was demonstrated that this compound was biodegraded in part via intermediates to tolylene-2,4-diamine.

INTRODUCTION

The biodegradability of polyurethanes has not previously been investigated in detail, but Darby et al. [1] have shown that polyester-based polyurethanes are susceptible to fungal attack. In this study, poly-D,L-lactic acid (PLA) polyurethane

251

compounds were synthesized, and the relationship between the biodegradability (rate of weight loss) of these polyurethanes and their polymer composition was investigated. The composition of polyurethane compounds before and after exposure to fungal attack was also analyzed by proton NMR; furthermore, measurement of oxygen consumption during cultivation of fungi was carried out to determine which fractions of the polyurethane compounds were biodegraded to CO_2. In addition, low molecular weight urethane compounds were synthesized and used to screen microorganisms capable of degrading urethane groups.

MATERIALS AND METHODS

Synthesis of Substrates

Synthesis of PLA Polyols

Poly-D,L-lactic acid (PLA) polyols were synthesized by reacting polyethylene glycol (PEG) and 2-chloropropionic acid sodium salt at 80–90°C in anhydrous DMF. The polyol consisted of PEG molecules with short terminal polylactic acid chains.

Synthesis of PLA Polyurethane Compounds

PLA polyurethane compounds were prepared by the reaction of the above polyols with polymethylene polyphenyl polyisocyanate (pMDI) and 0.3% triethylene diamine catalyst at 70°C in anhydrous conditions (Fig. 1a).

Synthesis of Urethane and Urea Model Compounds

Compound A (bisethylurethane of diphenylmethane diisocyanate) was synthesized by reacting diphenylmethane diisocyanate (MDI) with ethanol (Fig. 1b); Compound B (bisphenylurea of monomeric diphenylmethane diisocyanate) was

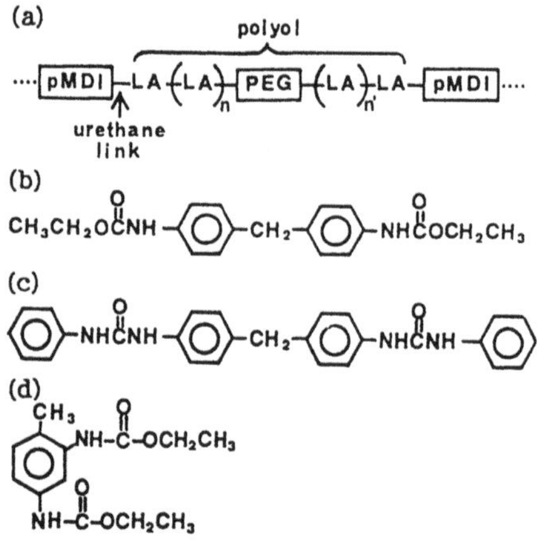

FIG. 1. Structures of (a) a poly-D,L-lactic acid polyurethane; (b) bisethylurethane of diphenylmethane diisocyanate, Compound A; (c) bisphenylurea of diphenylmethane diisocyanate, Compound B; (d) bisethyurethane of tolylene-2,4-diisocyanate, Compound C.

synthesized by reacting MDI with aniline (Fig. 1c) [2]; Compound C (bisethylure-thane of tolylene-2,4-diisocyanate) was synthesized by reacting tolylene-2,4-diisocyanate with ethanol (Fig. 1d). All these compounds were synthesized in anhydrous CH_2Cl_2 at 20°C.

Cultivation and Analysis

Measurement of Biodegradation (weight loss) of Polyurethane Compounds

Polyurethane compounds were incubated at 25°C and shaking at 130 rpm in a culture medium (NH_4NO_3 0.1%, K_2HPO_4 0.07%, KH_2PO_4 0.07%, $MgSO_4$ 0.07%, yeast extract 0.05%, NaCl 0.0005%, $ZnSO_4$ 0.0002%, $FeSO_4$ 0.0002%, $MnSO_4$ 0.0002%, pH 6.5) inoculated with spores of the following fungi: *Penicillium citrinum, P. funiculosum, Aspergillus niger, Cladosporum herbarum, Trichoderma* sp., *Rhizopus stolonifer*, and *Chaetosporum globosum* (as recommended by the ASTM). Before and after cultivation, the polyurethane samples were washed in distilled water, vacuum dried at 45°C for 5 hours, and weighed. The composition of polyurethanes was calculated from proton NMR spectra (Fig. 2).

Measurement of Biochemical Oxygen Consumption of Materials

The oxygen consumption during cultivation of fungi was measured in an inorganic salts medium [according to MITI (Ministry of International Trade and Industry, Japan) Method 301C], inoculated with spores of the above fungi.

Isolation of a Microorganism Capable of Biodegrading Compound C

A strain of fungus (Strain REN-11A) capable of biodegrading Compound C was isolated from soil obtained from a polyurethane factory after enrichment cultivation in an inorganic salts medium (Compound C 0.1%, K_2HPO_4 0.2%, KH_2PO_4 0.2% NH_4NO_3 0.1%, $MgSO_4$ 0.05%, Na_2SO_4 0.001%, $FeSO_4$ 0.0004%, $ZnSO_4$ 0.0004%, $MnSO_4$ 0.0002%, pH 6.2) at 27°C and shaking at 200 rpm.

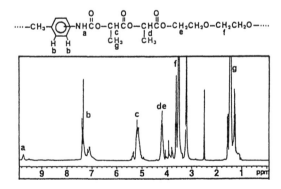

FIG. 2. ^{1}H-NMR spectrum of a PLA polyurethane (270 MHz; solvent, DMSO; 50°C).

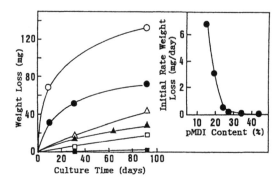

FIG. 3. Effect of the PMDI content on the weight loss of PLA polyurethanes (initial weight 200 mg). PMDI content: (○) 15.5%; (●) 19.0%; (△) 24.8%; (▲) 27.7%; (□) 34.4%; (■) 44.5%.

RESULTS AND DISCUSSION

Biodegradation of PLA Polyurethane Compounds

The rate of biodegradation (weight loss) in shaking fungal culture was less for PLA polyurethanes with higher pMDI content (Fig. 3). Polyurethanes prepared from polyols containing higher molecular weight PEG were biodegraded more rapidly than those made from polyols with lower molecular weight PEG (Fig. 4). The rate of biodegradation of polyurethanes decreased when the lactic acid content was increased (Fig. 5); however, a polyurethane compound made from a polyol containing no lactic acid (in other words, a polyethylene glycol polyurethane) was not biodegraded at all. The results of proton NMR analysis of the composition of polyurethanes before and after biodegradation are shown in Fig. 6; after 70 days the quantity of lactic acid and PEG in the polyurethane decreased by 55 and 74%, respectively, while the quantity of pMDI decreased by only 16%. This result suggests that the weight loss of the polyurethane compound is mainly due to biodegradation of the polyol segment within the polyurethane.

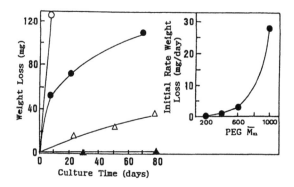

FIG. 4. Effect of the molecular weight of PEG in the polyol on the weight loss of PLA polyurethanes (initial weight 200 mg). PEG $\overline{M}_n$: (○) 1000; (●) 600; (▲) 400; (△) 200.

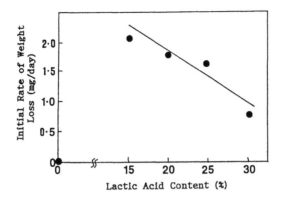

FIG. 5.   Effect of the lactic acid content on the weight loss of PLA polyurethanes (initial weight 200 mg; PEG $\overline{M}_n = 600$).

Oxygen Consumption During Cultivation of Fungi on a PLA Polyurethane, PLA, PEG, and Model Compounds

In order to determine which fractions of the polyurethane are biodegraded to CO_2, the oxygen consumption during cultivation of fungi on a PLA polyurethane, PEG, PLA, as well as Compounds A, B, and C was measured (Table 1). The polyurethane, PLA, and PEG were all readily biodegraded, while none of the urethane and urea compounds were biodegraded by the above ASTM fungi. Thus, only the polyol of PLA polyurethanes can be biodegraded to CO_2.

Biodegradation of Urethane Model Compound C

REN-11A was cultivated for 2 months in an inorganic salts medium with Compound C as the sole carbon source. After removal of the water of the culture medium by evaporation, the residual products were redissolved in CH_2Cl_2 and analyzed by GC-MS. Three compounds were detected by gas chromatography (Fig. 7). The mass spectrum of Peak 1 in the gas chromatogram (Fig. 8a) shows that this compound is likely to have been tolylene-2,4-diamine (TDA), and the mass spectra of Peaks 2 and 3 (Figs. 8b and 8c) show that these compounds could have been degradation intermediates of Compound C. The consumption of Compound C and

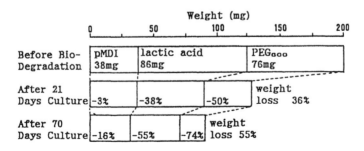

FIG. 6.   Composition of a PLA polyurethane before and after biodegradation.

TABLE 1. Oxygen Consumption during Cultivation of Fungi on a PLA Polyurethane, PLA, PEG, and Urethane and Urea Compounds[a]

Substrate	% Biodegradation after 30 days culture
Poly-D,L-lactic acid polyurethane	45
Polyethylene glycol ($\overline{M}_n$ 600)	36
Poly-D,L-lactic acid ($\overline{M}_n$ 1050)	70
Compound A:	1.5
Compound B:	0.3
Compound C:	0.8

Compound A:

$$CH_3CH_2O\overset{O}{\overset{\|}{C}}NH-\bigcirc-CH_2-\bigcirc-NH\overset{O}{\overset{\|}{C}}OCH_2CH_3$$

Compound B:

$$\bigcirc-NH\overset{O}{\overset{\|}{C}}NH-\bigcirc-CH_2-\bigcirc-NH\overset{O}{\overset{\|}{C}}NH-\bigcirc$$

Compound C:

$$NH-\overset{O}{\overset{\|}{C}}-OCH_2CH_3$$
$$NH-\overset{O}{\overset{\|}{C}}-OCH_2CH_3$$

[a]Measured according to MITI (Ministry of International Trade and Industry, Japan)
Method 301C.

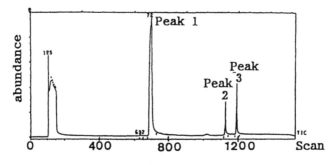

FIG. 7.   Gas chromatogram of the products of degradation of Compound C by fungal
strain REN-11A (column, OV-17, 0.247 mm × 30 m × 0.25 μm; carrier gas (He) flow rate,
45 mL/min).

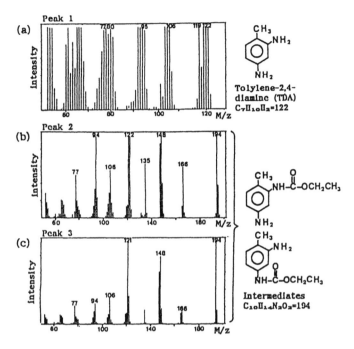

FIG. 8.   Mass spectra of the compounds detected in the gas chromatogram of Fig. 7
(ion voltage, 70 eV; ion source temperature, 240°C).

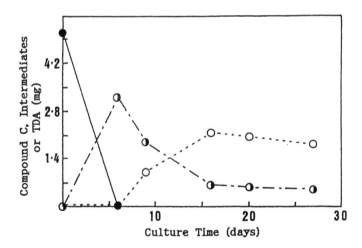

FIG. 9.   Consumption of Compound C and accumulation of degradation products by
fungal strain REN-11A (HPLC conditions: column, ODS 3C18, 4 × 50 mm; mobile phase,
acetonitrile–water (50:50 v/v), flow rate 0.5 mL/min; UV detection at 225 nm). (●) Compound C; (◐) intermediates; (○) tolylene-2,4-diamine (TDA).

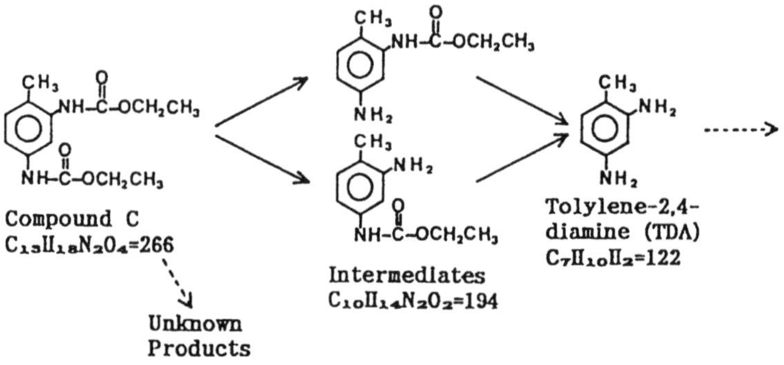

FIG. 10. Postulated pathway for degradation of bisethylurethane of diphenylmethane diisocyanate (Compound C).

the accumulation of the intermediates and TDA over time were analyzed by HPLC (Fig. 9). Compound C was completely consumed within 6 days of culture, and about 62% of Compound C accumulated as intermediates. Between 6 and 16 days of culture, the amount of intermediates decreased; the amount of accumulated TDA roughly corresponded to the amount of intermediates consumed. After 16 days, the amount of TDA decreased slightly. It seems likely, therefore, that degradation of Compound C occurs according to the scheme shown in Fig. 10.

CONCLUSION

The rate of weight loss of poly-D,L-lactic acid (PLA) polyurethane compounds in shaking fungal culture increased when: 1) the polymethylene polyphenyl polyisocyanate (pMDI) content of the polyurethane was reduced; 2) the molecular weight of the polyethylene glycol (PEG) moiety in the polyol was increased; 3) the lactic acid content of the polyol was increased. Proton NMR analysis revealed that the weight loss of the polyurethane was mainly due to biodegradation of the PEG and lactic acid of the polyol. Measurement of oxygen consumption during cultivation of fungi indicated that only the PEG and PLA fractions of the polyol in PLA polyurethanes were biodegraded to CO_2, while the pMDI fraction was not biodegraded.

A strain of microorganism capable of biodegrading a urethane compound (bisethylurethane of tolylene-2,4-diisocyanate) was isolated, and it was demonstrated that this compound was biodegraded in part via intermediates to tolylene-2,4-diamine (TDA). Therefore, there is a possibility that urethane groups within polyurethanes can be biodegraded.

REFERENCES

[1]   R. T. Darby and A. M. Kaplan, *Appl. Microbiol.*, *16*, 900 (1968).
[2]   T. M. Chapman, *J. Polym. Sci., Polym. Chem. Ed.*, *27*, 1993 (1989).

BIODEGRADATION OF POLYESTER COPOLYMERS CONTAINING AROMATIC COMPOUNDS

U. WITT, R.-J. MÜLLER, and W.-D. DECKWER

Biochemical Engineering
GBF, Gesellschaft für Biotechnologische Forschung mbH
Mascheroder Weg 1, D-38124 Braunschweig, Germany

ABSTRACT

For investigation of the microbial accessibility of polyesters based on 1,3-propanediol, a series of different polymer structures (homo, random, and block copolymers) were synthesized by polycondensation of terephthalic acid, adipic acid, sebacic acid, and 1,3-propanediol. The alcohol component, 1,3-propanediol, can be obtained from a biotechnological process from glycerol, a surplus product of the oleochemical industry. Aliphatic dicarbonic acids can be derived from vegetable oils. Biodegradation was performed in different test systems. 1) Polyester films were exposed to an aerated liquid medium inoculated with eluates from soil. 2) For this test system, polymer films were buried in soil. Copolyesters exhibited significant differences in both tests. Furthermore, a clear influence of the polymeric structure as well as of the chain length of the aliphatic dicarbonic acids on the microbial accessibility was observed.

INTRODUCTION

Due to increasing problems in waste management, biodegradable polymers have become of new interest. In particular, a combination of biodegradability and the application of renewable sources offers the chance to make plastics part of natural cycles. Biologically synthesized and degradable polymers such as PHB show very specific properties; however, these are of limited variability. Preparation of

259

plastics from different monomers, on the other hand, offers the possibility to create biodegradable polymers with tailor-made properties. At GBF, a series of polyesters based on renewable sources has been synthesized [1]. For example, 1,3-propanediol, exclusively used as a diol component, is obtained from a biotechnological process using glycerol, a surplus product of the oleochemical industry (>1 million tons per year) [2–4]. Aliphatic dicarbonic acids can be derived from vegetable oils; for instance, by ozonolysis [5]. These types of polyesters, biodegradable through cleavage of ester bonds by hydrolysis, are especially appropriate to detect correlations between the structure of polyesters and their biodegradability.

BIOLOGICAL TEST SYSTEMS

In order to investigate the microbial accessibility of such polyesters, different polymer structures (homo, random, and block copolymers) were synthesized [1]. The degradation tests were carried out with polyester films of 100 μm thickness by 25 mm in diameter. The films were exposed to an aerated liquid medium at 30°C for 8 weeks. This medium contained only mineral salts to guarantee appropriate physiological conditions. Test media were inoculated with 1% (v/v) eluate from soil. Sterile blank tests were carried out to evaluate the resistance of the polymers against (chemical) hydrolysis. Figure 1 illustrates the correlation between melting temperatures (T_m) and weight losses after biodegradation of synthesized aliphatic polyesters based on 1,3-propanediol. A decrease of weight loss with increasing melting points of aliphatic dicarbonic acids was observed for the homopolyesters synthesized from dicarbonic acids with a carbon chain length larger than 2. However, these homopolyesters showed only limited applicability (brittle films, low melting temperatures), even though weight-average molecular weights were in the range of technical relevant polycondensates (approximately 25,000 g/mol). As an exception, the aliphatic polyester with an oxalic acid as the dicarbonic component showed a surprisingly high melting point and good mechanical properties. In this case, further investigations with regard to the influence of chemical hydrolysis are in preparation. In order to improve the poor material properties of aliphatic polyes-

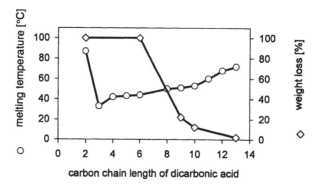

FIG. 1.   Melting points and weight losses of aliphatic dicarbonic acids based on 1,3-propanediol.

ters, statistical as well as block copolyesters were synthesized by polycondensation of 1,3-*propane*diol, *terephthalic* acid dimethyl ester, and the dimethylesters of aliphatic dicarbonic acid such as (*sebacic* acid and *adipic* acid, respectively). The melting temperatures of the statistical polyesters with *sebacic* acid as the aliphatic acid component (hereinafter called PTS) and the weight losses of the films after degradation are shown to depend on the composition in Fig. 2(a). Corresponding to the aliphatic materials, the rate of degradation decreases with an increase in melting temperatures. Indeed, from values of the molar fraction approximately larger than 0.3, the melting points and the mechanical properties drastically increased and no weight loss for the statistical polyester PTS 44:56 (molar ratio of *terephthalic* acid:*sebacic* acid) was observed. For a second set of statistical polyesters containing *adipic* acid as the aliphatic acid component (hereinafter called PTA), the same behavior with respect to the correlation between T_m and weight loss after biodegradation was observed (see Fig. 2b). In contrast to the statistical copolyester PTS 44:56, a block copolyester having a similar ratio of aromatic and aliphatic acid (48 mol% terephthalic and 52 mol% sebacic acid) showed a significant weight loss although the T_m (131°C) is comparable with that of the statistical polyester PTS 44:56, $T_m = 139$°C (see Fig. 3). Therefore, the clear influence of

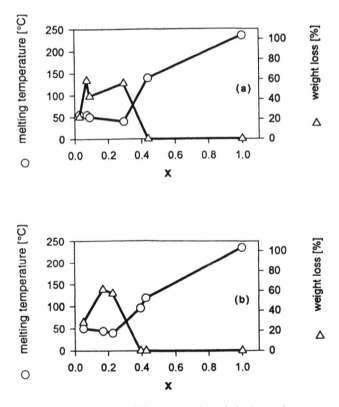

FIG. 2.   Melting points and weight losses of statistical copolyesters containing 1,3-propanediol, terephthalic acid, sebacic acid (a), and adipic acid (b), respectively, as a function of molar fraction of terephthalic acid in the copolyester.

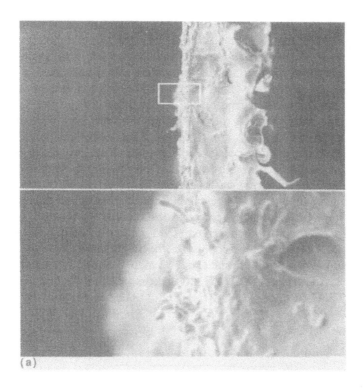

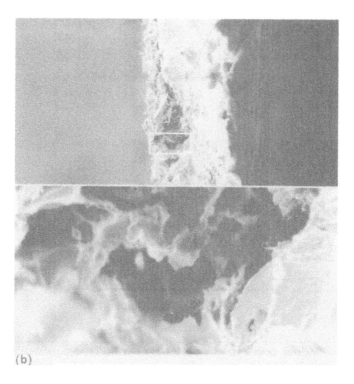

FIG. 3.   SEM micrographs of a block copolyester film: (a) original film; (b) film after 4 weeks exposure to a liquid medium containing microorganisms from soil. Enlargement: above, 500-fold;·below, 4000-fold.

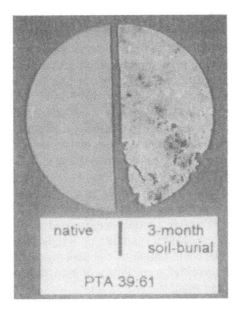

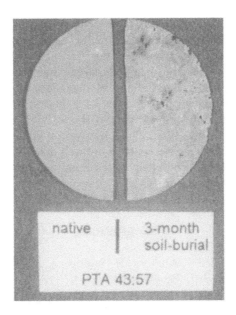

FIG. 4.   Photographs of statistical copolyester films PTA 39:61 and PTA 43:57 (native = original films).

polymeric structure on biological resistance becomes evident upon comparison of statistical and block copolyester, both exhibiting similar melting points and similar stoichiometric compositions. These findings show what type of polymer with regard to its sequential structure predominates biodegradability (whether yes or no) whereas the physical properties (e.g., T_m) of the polymers obviously affect the rate of an existing biodegradation (see also Fig. 1). Further investigations will concentrate on the question whether or not only the long aliphatic chains in the block copolyester are exclusively biodegraded and what the fate of the aromatic blocks is. Are they eventually left as a resistant biodegradation residue?

As described before, statistical polyester films with a molar fraction of terephthalic acid approximately larger than 0.3 showed no degradation in the liquid test media. In contrast, the polyester films buried in soil showed a significant weight loss (approximately 20%) for the statistical polyester PTA 39:61 (T_m = 96°C) and PTA 43:57 (T_m = 119°C) after 3-month exposure (Fig. 4), whereas the statistical polyester PTS 44:56 (T_m = 139°C) exhibits no weight loss after soil burial over 6 month. Furthermore, the chain length of the aliphatic dicarbonic acid had an evident influence on the microbial accessibility of statistical copolyesters which have similar melting points and stoichiometric compositions.

CONCLUSION

This study revealed a dependence of biological accessibility on the sequential structure of polyester copolycondensates. For statistical copolyesters with a molar fraction of terephthalic acid approximately larger than 0.3, an adjustment of the optimum between physical properties and biodegradability appears to be feasible. Thus, controlling the biological properties of a plastic material by its chemical structure is a promising subject in the final design of biodegradable polymers. The significant differences in biodegradation observed by application of two different degradation tests elucidates the necessity to use well-defined laboratory tests as well as simulation tests. The latter will help to select natural degradation conditions for a final judgment of biodegradability.

REFERENCES

[1]   U. Witt, R.-J. Müller, J. Augusta, H. Widdecke, and W.-D. Deckwer, *Macromol. Chem. Phys., 195*, 793 (1994).
[2]   Th. Homann, C. Tag, H. Biebl, W.-D. Deckwer, and B. Schink, *Appl. Microbiol. Biotechnol., 33*, 121 (1990).
[3]   H. Biebl, S. Marten, H. Hippe, and W.-D. Deckwer, *Ibid., 36*, 592 (1992).
[4]   B. Günzel, S. Yonsel, and W.-D. Deckwer, *Ibid., 36*, 289 (1991).
[5]   H. Baumann, M. Bühler, H. Fochen, F. Hirsinger, H. Zoebelein, and J. Falbe, *Angew. Chem., 100*, 41 (1988).

PROPERTIES AND POLYMERIZATION OF BIODEGRADABLE THERMOPLASTIC POLY(ESTER-URETHANE)

MIKA HÄRKÖNEN, KARI HILTUNEN, MINNA MALIN, and JUKKA V. SEPPÄLÄ

Department of Chemical Engineering
Helsinki University of Technology
Kemistintie 1, 02150 Espoo, Finland

ABSTRACT

Aliphatic polyesters, such as poly(lactic acids), need high molecular weight for acceptable mechanical properties. This can be achieved through ring-opening polymerization of lactides. The lactide route is, however, relatively complicated, and alternative polymerization routes are of interest. In this paper we report the properties of a polymer made by a two-step process: first a condensation polymerization of lactic acid and then an increase of the molecular weight with diisocyanate. The end product is then a thermoplastic poly(ester-urethane). The hydroxyl-terminated prepolymer was made with condensation polymerization of L-lactic acid and a small amount of 1,4-butanediol. The polymerization was performed in the melt under nitrogen and reduced pressure. The preparation of poly(ester-urethane) was done in the melt using aliphatic diisocyanates as the chain extenders reacting with the end groups of the prepolymer. The polymer samples were carefully characterized, including preliminary degradation studies. The results indicate that this route to convert lactic acid into thermoplastic biodegradable polymer has high potential. Lactic acid is converted into a mechanically attractive polymer with high yield, which could make the polymer suitable for high volume applications. The mechanical properties of the poly(ester-urethane) are comparable with those of poly(lactides). Capillary rheometer measurements indicate that the polymer is processible both by injection molding and extrusion.

INTRODUCTION

It has become a rather widely adopted opinion that biodegradable polymers have a well-grounded role in solving the waste problem. When composting of waste becomes more common, normal "stable" thermoplastics can cause problems in some applications, especially when those plastics are combined with otherwise compostable material. However, at present the relatively poor properties and high price of compostable plastics make their utilization in volume applications unattractive and practically impossible.

The feasibility of production processes and the possibility to use renewable raw materials have made lactic acid an attractive monomer for biodegradable polymers. However, poly(lactic acid) needs a relatively high molecular weight to have acceptable mechanical properties in many applications [1]. Conventional condensation polymerization of lactic acid does not increase the molecular weight enough, and the yield of cyclic side products, such as lactides, is often too high. Acceptable molecular weights can be achieved through ring-opening polymerization of lactides. This route is, however, relatively complicated due to the multistep process from lactic acid through a lactide intermediate into the final poly(lactide), which makes the total yield rather low. Therefore, alternative polymerization routes are of interest.

In this paper, properties of a biodegradable thermoplastic poly(ester-urethane) based on lactic acid are reported. The polymer is produced by a two-step process: first, a condensation polymerization of lactic acid and then an increase of the molecular weight with diisocyanate. In the literature there have been only a few publications [2–4] where resembling polymers have been reported. For example, Hori et al. [2] published an article describing the synthesis of poly(ester-urethane) from poly(3-hydroxybutyrate) segments.

EXPERIMENTAL

In the preparation of poly(L-lactic acid) prepolymer, a mixture containing 98 mol% L-lactic acid and 2 mol% 1,4-butanediol was condensation polymerized in a rotation evaporator under a nitrogen stream and reduced pressure. Stannous octoate was used as a catalyst.

The poly(ester-urethane) was prepared from 250 g prepolymer and 20 mL 1.6-hexamethylene diisocyanate in a stirred glass reactor in melt under a nitrogen atmosphere.

The molecular weights were analyzed with GPC (Waters) at room temperature using polystyrene standards. In addition, the poly(L-lactic acid) prepolymer was characterized with ^{13}C NMR (Varian Unity 400, working at 100,577 MHz) to quantitatively determine the amount of hydroxyl end groups, lactic acid, and lactide residues. FT-IR analysis were done with a Nicolet Magna-IR Spectrometry 750. The mechanical properties were determined from compression-molded specimens with an Instron 4204. The DSC analysis were made by a Polymer Laboratories DSC and DMTA analysis by a Perkin-Elmer DMA 7. Rheological measurements were done with a capillary rheometer (Göttfert Rheograph) having a die of 20:1.

RESULTS AND DISCUSSION

The poly(L-lactic acid) prepolymer had a number-average molecular weight of 4500 g/mol measured with GPC. The prepolymer had 5.7 wt% L-lactide and 5.6 wt% L-lactic acid residues measured with ^{13}C NMR. The hydroxyl termination of the prepolymer was confirmed and quantitatively measured with ^{13}C NMR. The number-average molecular weight calculated from the ratio between main chain carbons and chain ends was 2100 g/mol.

The product poly(ester-urethane) had a number-average molecular weight of 32,000 g/mol (with GPC), and according to FT-IR analysis it did not contain any free isocyanate groups. The polymer contained 1.6 wt% L-lactide residue, but no L-lactic acid was detected. The total conversion of L-lactic acid to poly(ester-urethane) was about 96 mol%.

Mechanical properties of this poly(ester-urethane) (PEU) compared to some other relevant polymers are seen in Fig. 1. The tensile strength and modulus of PEU are very close to the corresponding values of poly(L-lactide) made by the

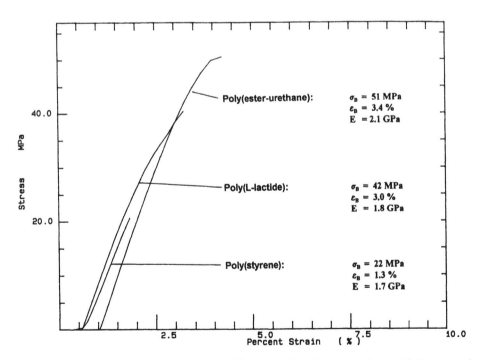

FIG. 1.   Stress–strain curves and tensile properties of compression-molded test specimens of the poly(ester-urethane) and some reference materials.

ring-opening route. Typically these aliphatic polyesters resemble polystyrene, although polystyrene is more brittle.

DSC analysis shows that PEU is a totally amorphous polymer having a clear glass transition temperature (T_g) at 53°C. The three-point bending DMTA analysis in Fig. 2 shows that the storage modulus drastically decreases at a temperature of about 50°C, and that the T_g measured from the peak value of tan δ is 62°C. Thermal analysis indicates that PEU loses its mechanical properties at temperatures above 50°C.

Capillary rheology analysis (Fig. 3) shows that PEU is thermoplastic. It is processible by conventional methods. The melt viscosity is suitable for injection molding at a temperature of 210°C in typical shear rates of injection molding (2000 L/s). At 180°C the melt viscosity is at an acceptable level for extrusion. Thermal degradation may be partially responsible for the marked decrease in the melt viscosity between 180 and 210°C.

According to degradation studies of poly(urethane) mentioned in the literature [5, 6], it can be assumed that degradation behavior is relatively close to the corresponding properties of the prepolymer used. The preliminary hydrolysis test series showed that poly(ester-urethane) is hydrolytically degradable, though the degradation seems to be slower than that of poly(L-lactide). For example, the number-average molecular weight of a PEU sample decreased during hydrolysis at 37°C (pH 7) from an initial 49,000 g/mol to 43,000 g/mol in 1 week and to 5,000 g/mol after 8 weeks (Figure 4). The degradation behavior resembles that of poly(L-lactide) [7], and practically no weight loss was observed during the test period. However, these hydrolysis tests were only preliminary, and no final conclusions about degradation can be made.

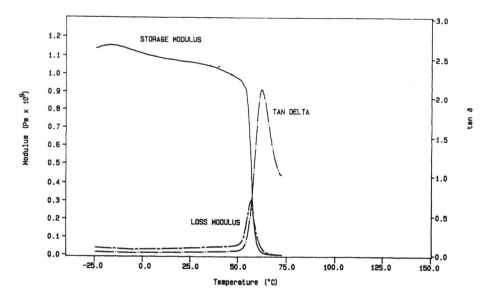

FIG. 2.   The three-point bending DMTA analysis of the poly(ester-urethane).

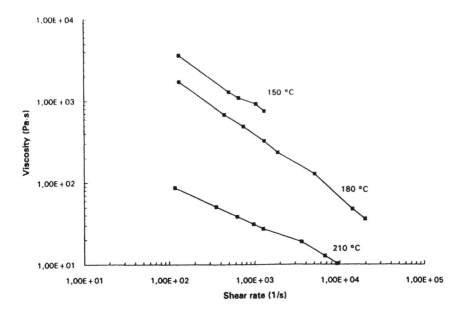

FIG. 3.   Capillary rheometer analysis of the poly(ester-urethane).

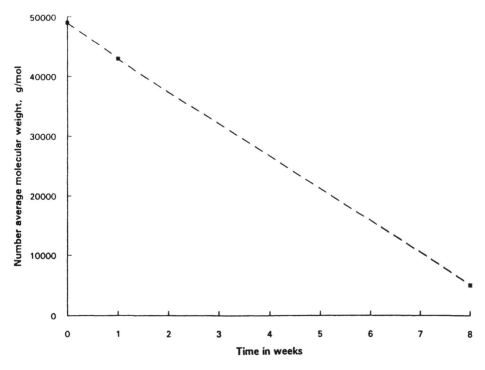

FIG. 4.   Decrease of the number-average molecular weight of the poly(ester-urethane) during hydrolysis at 37°C and pH 7.

CONCLUSIONS

According to this study, this polymerization route to a thermoplastic biodegradable polymer based on lactic acid has high potential. Lactic acid can be converted into a mechanically attractive polymer with high yield (>95%), and the polymerization is technically feasible, at least on a laboratory scale.

The mechanical properties of this poly(ester-urethane) are comparable to those of poly(lactide)s made through a ring-opening route. The rheological measurements indicate that the polymer is processible with conventional methods such as injection molding and extrusion. The polymer loses its stiffness at about 50°C, which can be too low for many applications. The preliminary degradation studies indicate that the polymer is hydrolytically degradable.

REFERENCES

[1]   I. Engelberg and J. Kohn, *Biomaterials, 12*, 292 (1991).
[2]   Y. Hori, M. Suzuki, Y. Okeda, T. Imai, M. Sakaguchi, Y. Takahaski, A. Yamaguchi, and S. Akutagawa, *Macromolecules, 25*, 5117 (1992).
[3]   H. Torii, "Biodegradable Thermoplastic Polyurethanes," Regurusu K.K. JP 04013710 A2, (1992).
[4]   Y. Doi and E. Takyama, "Preparation of Urethane Bond-Containing Biodegradable Polylactides" (Showa Highpolmer Co. Ltd.), JP 05148352 A2 (1993).
[5]   S. J. Huang, C. Marci, M. Roby, C. Benedict, and J. A. Cameron, *ACS Symp. Ser., 62*, 471 (1981).
[6]   Y. Tokiwa, T. Ando, T. Suzuki, and T. Takeda, *Polym. Mater. Sci. Eng., 62*, 988 (1990).
[7]   J. V. Seppälä and R. M. H. Malin, "Studies of the Hydrolytic Degradation of Some Aliphatic Polyesters," in *3rd International Workshop on Biodegradable Plastics and Polymers,* Osaka, Japan, November 9–11, 1993.

BIODEGRADATION OF STARCH-BASED MATERIALS

MINNA VIKMAN, MERJA ITÄVAARA, and KAISA POUTANEN

VTT Biotechnology and Food Research
Biologinkuja 1, P.O. Box 1500, FIN-02044 VTT, Finland

ABSTRACT

The aim of this study was to compare three different test methods for assaying the biodegradability of starch-based materials. The materials tested included some commercial starch-based materials and thermoplastic starch film prepared by extrusion from native potato starch and glycerol. Enzymatic hydrolysis was performed using excess *Bacillus licheniformis* α-amylase and *Aspergillus niger* glucoamylase at 37°C. The degree of degradation was assayed by measuring the dissolved carbohydrates and the weight loss of the samples. The head-space test was based on carbon dioxide evolution using sewage sludge as an inoculum. The composting experiments were carried out in an insulated commercial composter bin. The degradation was evaluated visually at weekly intervals, and the weight loss of the samples was measured after composting. Good correlation was found among the three different test methods.

INTRODUCTION

The biodegradation of starch-based materials depends on the starch processing method used as well as on the biodegradability of other components. The main elements in biodegradability testing are incubation of the sample under conditions conducive to microbial attack and/or their enzymes, and evaluation of the degree of degradation [1]. Several studies have been carried out to evaluate biodegradation by incubating samples in a compost environment [2, 3] or measuring the carbon dioxide evolution in aquatic conditions [4, 5]. Using specific enzyme assays, the time needed to perform the biodegradation test can be reduced. According to some

271

studies [6, 7], the ability of amylolytic enzymes to degrade starch composites can be rather limited.

We are currently studying the biodegradation of starch-based materials by several biodegradation test methods at VTT Biotechnology and Food Research. These tests include methods based on carbon dioxide evolution, composting, and enzymatic testing. Some preliminary results obtained are reported in this paper.

MATERIALS AND METHODS

Materials

Biopac was obtained from Austrian Biologische Verpackungssysteme GmbH (Sample 1) and Biopur from Biotec GmbH & Co. KG (Sample 2). Mater-Bi ZF02U film (Sample 3) was from Novamont North America. Thermoplastic starch film was prepared by extrusion from glycerol and native potato starch (Sample 4).

Test Methods

Enzymatic hydrolysis was performed at 37°C. A 100-mg sample was incubated with excess *Bacillus licheniformis* (Genencor International Europe Ltd., Finland), α-amylase, and *Aspergillus niger* glucoamylase (Boehringer Mannheim, Germany) in 10 mL of 0.1 M acetate buffer, pH 5.0. The degree of degradation during 24 hours was assayed by measuring the dissolved carbohydrates [8] and the weight loss of the samples.

The head-space test was based on carbon dioxide evolution using sewage sludge as an inoculum. A 30-mg sample was incubated with the sewage sludge microorganisms at 25°C in 50 mL mineral nutrient solution (ASTM D5247-92), and carbon dioxide evolution was measured at weekly intervals.

The composting experiments were carried out in an insulated commercial composter bin (Sepe, Suomen kompostointipalvelu Ky, Finland) filled with biowaste. The composting test was based on naturally occurring composting reactions, and no external heating and ventilation was employed. Square sheets of tested materials were attached to the steel frame which was buried in the biomass. The degradation was evaluated visually at weekly intervals, and the weight loss was measured after composting. To ensure that the composting proceeded as expected, temperature, pH, moisture, and oxygen and carbon dioxide concentrations inside the composter bin were measured.

RESULTS

During the enzymatic hydrolysis the materials tested were hydrolyzed extensively with the exception of Sample 3, in which starch was blended with other polymers (Fig. 1). The weight loss of the samples was larger than the amount of solubilized carbohydrates, indicating the solubility of other components besides starch. The enzymatic method used represents a rapid means of obtaining preliminary information about the biodegradability of starch-based materials.

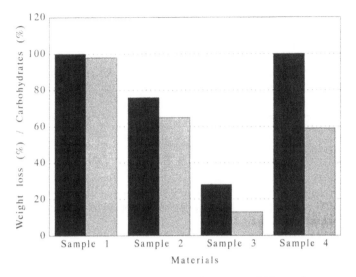

FIG. 1.   Weight loss and dissolved carbohydrates after 24 hours in enzymatic hydrolysis. ■: weight loss (%); ▨: dissolved carbohydrates (% dry weight).

In the head-space test the highest carbon dioxide evolution was measured
during the first 2 weeks of incubation (Fig. 2). The largest CO_2 evolution (%
theoretical) was obtained when Samples 1 and 4 were tested. The advantages of the
head-space test are evident: it is easy to perform and a large number of samples can
easily be tested.

Samples 1 and 4 were degraded completely after 49 days of composting (Table
1). The weight loss of Sample 2 was 65%, and its thickness was reduced but no

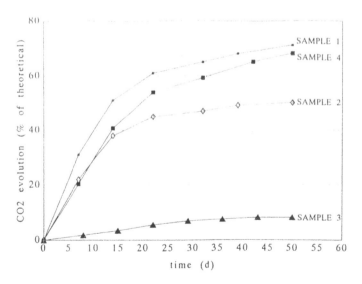

FIG. 2.   Carbon dioxide evolution (% theoretical) in the head-space test.

TABLE 1.   Biodegradation of the Samples by the Three Different
Test Methods

Material	Enzymatic test, 24 hours, weight loss, %	Head-space test, 50 days, CO_2 evolution,[a] %	Compost test, 49 days, weight loss, %
Sample 1	100	100	100
Sample 2	75	73	65
Sample 3	28	14	45
Sample 4	100	100	100

[a]CO_2 evolution of the sample (% theoretical)/CO_2 evolution of the
native starch (% theoretical).

holes were formed during composting. Sample 3 became very brittle and the weight
loss was 45%. The disadvantage of the composting test is that it is time consuming.
It is also difficult to control the composting parameters, and the reproducibility of
the results is poorer than with the two other methods.

CONCLUSION

Good correlation was found among the three different test methods. In the
case of blended polymers such as Sample 3, the most excessive degradation was
obtained by the composting test. The conditions in the compost environment are
ideal for degradation mainly because of the mixed microbial population, the broad
diversity of the enzymes secreted, and the high temperature during the thermophilic
phase of composting.

REFERENCES

[1]   T. M. Aminabhavi, R. H. Balundgi, and P. E. Cassidy, *Polym. Plast. Technol. Eng., 29*, 235 (1990).
[2]   D. F. Gilmore, S. Antoun, R. W. Lenz, S. Goodwin, R. Austin, and R. C. Fuller, *J. Ind. Microbiol., 10*, 199 (1992).
[3]   H. B. Greizerstein, J. A. Syracuse, and P. A. Kostyniak, *Polym. Degrad. Stab., 39*, 251 (1993).
[4]   J. Struijs and J. Stoltenkamp, *Exotoxicol. Environ. Saf., 19*, 204 (1990).
[5]   R. R. Birch and R. J. Fletcher, *Chemosphere, 23*, 507 (1991).
[6]   P. Allenza, J. Schollmeyer and R. P. Rohrbach, in *Degradable Materials* (S. A. Barenberg, J. L. Brash, R. Narayan, and A. E. Redpath, Eds.), CRC Press, Boca Raton, Florida, 1990, p. 357.
[7]   J. M. Gould, S. H. Gordon, L. B. Dexter, and C. L. Swanson, *ACS Symp. Ser., 433*, 65 (1990).
[8]   M. Dubois, K. A. Gilles, J. K. Hamilton, P. A. Rebers, and F. Smith, *Anal. Chem., 28*, 350 (1956).

BIODEGRADATION OF POLYCAPROLACTONE BY MICROORGANISMS FROM AN INDUSTRIAL COMPOST OF HOUSEHOLD REFUSE. PART II

F. LEFEBVRE, A. DARO, and C. DAVID

Chimie des Polymères et Systèmes Organisés
Université Libre de Bruxelles
CP 206/1, Bld du Triomphe, 1050 Bruxelles, Belgium

ABSTRACT

The rate of biodegradation of polycaprolactone samples, hydroxy (M_n = 4000 and 37000) and methoxy (M_n = 4000) terminated, by mixed cultures of microorganisms from a suspension of compost, and by a pure culture of an actinomycete isolated from it, has been monitored by the measurement of oxygen consumption. The dependence on polymer molecular weight suggests that initiation of degradation occurs in the vicinity of chain ends. Further analysis of the residues allows primary considerations about the bacterial metabolism of the substrate.

INTRODUCTION

Biodegradation of two molecular weight (MW) samples of polycaprolactone (PCL) has been monitored by a respirometric method. Oxygen consumed by aerobic bacteria has been measured, allowing a kinetic approach of polymer degradation, by microorganisms from a suspension of compost, and by a pure culture of an actinomycete isolated from it. Further characterization includes mass loss evaluation of the polymer, gel permeation chromatographic (GPC) analysis of the residue, and the measurement of the biomass produced by microorganisms.

275

MATERIALS AND METHODS

Details on materials and methods have been described previously [1].

Polymers

Hydroxy-terminated PCL samples were used (M_n = 4,000 and 37,000). Purified samples and methoxy-terminated PCL4000 were prepared. The polymers were used as cut pieces of films (about 50 μm thick) for PCL37000 and as small film fragments for pure and methylated PCL4000 at a concentration of 0.2% (w/v) after sterilization. Crystallinity of the samples was determined by DSC using the value of 135.6 J/g for the melting enthalpy of a 100% crystalline sample [2].

Inoculum

Microorganisms from a suspension of an industrial compost of household refuse and a pure culture an actinomycete isolated from it were used.

Measurement of O_2 Consumption

The method consists in the manometric measurement of O_2 depletion from a closed bottle containing the polymer as carbon source and an inoculum of microorganisms in an inorganic basal medium at 35°C and without light (see Fig. 1).

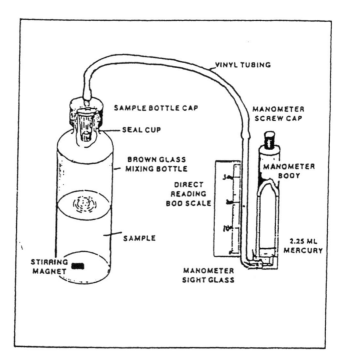

FIG. 1.   Manometric system (the polycaprolactone samples used in our tests are not soluble in the inorganic basal medium).

Released CO_2 is trapped by NaOH pellets in the seal cup of the bottle. A blank is run that contains only the inoculum and the inorganic basal medium. The O_2 uptake is subtracted from the value obtained with the test bottle. The difference is assigned to degradation of the carbon source [3]. O_2 consumption is expressed in percent of the theoretical value required to transform quantitatively the substrate in CO_2 and H_2O.

Further Analysis

The experiment was stopped at the plateau of O_2 consumption. The solution was filtered through a 5-μm Teflon filter. The residual polymer and the formed biomass are retained on the filter. Residual polymer is separated from biomass by quantitative extraction with CH_2Cl_2. Undegraded polymer, and hence weight loss, is determined after evaporation of the solvent. The filter is then dried and the biomass weight is measured. The residual polymer is analyzed by a GPC Waters 150C model with Ultrasyragel columns and compared with initial PCL samples, the same quantity being injected on the top of the columns in all cases; a calibration has been achieved with PCL standards.

RESULTS

Typical results are given in Table 1 and Fig. 2 for the first inoculum. Duplicates are reported in Table 1 and Fig. 3 for the actinomycete strain.

Microorganisms from a Suspension of Compost

A small amount of yeast extract (0.01%) was necessary to achieve biodegradation [1]. High O_2 consumption values and nearly complete degradation of the polymer samples were then obtained. More time was required, however, to reach the plateau for PCL37000. Biomass amounts are similar, but relatively low.

Pure Culture of the Actinomycete

Samples of purified PCL4000 and PCL37000 and of methylated PCL4000 were used without any addition of another C source. High values of O_2 consumption rates were registered for all of the tested samples, as well as nearly complete disappearance of the polymers. However, the polymer biodegradation requires different quantities of O_2 at the plateau and results in different amounts of biomass; PCL4000 and PCL37000 samples are differentiated through the O_2 consumption rates.

GPC Analyses

Residual PCL37000 of each test series (Fig. 4), as well as methoxy- or hydroxy-terminated PCL4000 (not shown) of the pure culture, were analyzed by GPC. In all cases, biodegradation of the polymers results in the disappearance or decrease of the high MW peak without any shift, but two small peaks appeared in

TABLE 1.   Biodegradation of Polycaprolactone

Inoculum	Substrate[a]	Substrate cristallinity, %	Oxygen consumption,[b] %	Weight loss, %	Biomass, mg	Carbon conversion to biomass,[d] %
Microorganisms from a suspension of compost	PCL 4000 + YE[c] (1)	76	53.8	89.4	17 ± 1	17 ± 1
	PCL 4000 + YE (2)		66.3	93.5		
	PCL 37000 + YE	55	78.0	93.9	16	16
Pure culture of actinomycete	PCL 4000(purified)	77	51.8	99.6	43 ± 4	39 ± 3
	PCL 4000(purified)		55.5	99.0		
	PCL 4000(methylated)	65	57.7	97.9	31 ± 10	29 ± 9
	PCL 4000(methylated)		61.7	93.9		
	PCL 37000(purified)	63	50.5	96.7	49 ± 3	44 ± 4
	PCL 37000(purified)		45.5	97.5		

[a]0.2% (w/v) in the inorganic basal medium.

[b]In percent of the theoretical value of O_2 consumption for complete transformation of the substrate into H_2O and CO_2.

[c]YE: yeast extract (Difco) 0.01% in the inorganic basal medium.

[d]Expressed as C biomass produced (mg)/C substrate used (mg) (C content in the biomass is estimated by the general formula $CH_{1.8}O_{0.5}NO_2$, which is a good average for many microorganisms) [4].

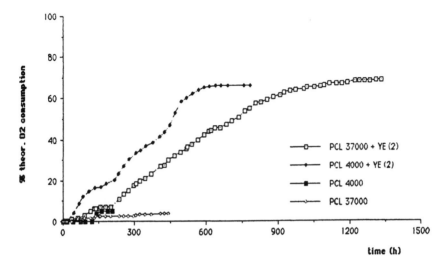

FIG. 2. O_2 consumption by a suspension of microorganisms from compost growing on polycaprolactone 0.2% in the basal medium. YE: yeast extract (0.01%) (Difco) in the basal medium.

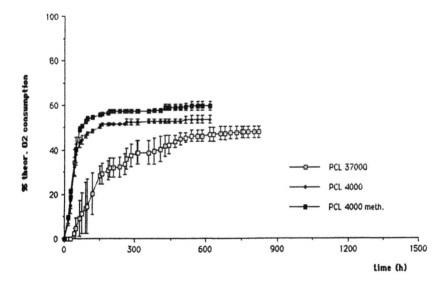

FIG. 3. O_2 consumption by a pure strain of an actinomycete isolated from compost growing on polycaprolactone 0.2% in the basal medium. The points shown are the mean values of the data from Table 1.

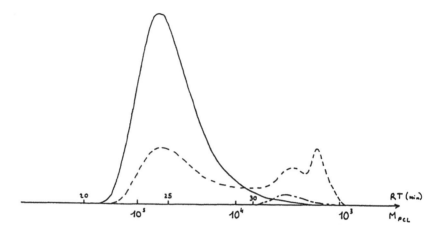

FIG. 4.   GPC diagram of PCL37000. (——) Initial polymer. (- - -) Residual polymer
after degradation by microorganisms from compost (6.1% residue). (- - -) Residual polymer
after degradation by a pure culture of actinomycete (3.3% residue).

the low MW area; they correspond to oligomers of M_n = 2930 and 1430. These
oligomers are present in very small amounts, since residual polymer is only a few
percent of the initial quantity. They are supposed to be degradation intermediates
that would finally be completely degraded.

DISCUSSION

Oxygen Consumption and Degradation Mechanisms

For both kinds of inocula, the rate of O_2 consumption by high molecular
weight polymer is always lower, and more time is required to reach the plateau. This
leads us to consider the degradation mechanism as an end-initiated unzipping process (see Ref. 1 for further details). This is not incompatible with the same O_2
consumption rates observed for HO- and CH_3O-terminated PCL4000. Indeed, the
methoxy-termination is at least five CH_2 units away from ester bonds, so it should
not inhibit the binding of the hydrolase enzyme to these bonds. Crystallinity does
not seem to retard biodegradation, the O_2 consumption rate being higher for the
more crystalline sample (PCL4000) than for the less crystalline one (PCL37000).

Oxygen Consumption and Carbon Conversion to Biomass

Upon biodegradation, the carbon source is transformed into biomass, CO_2,
and eventually secondary products. Oxidation to CO_2 results mainly from the tricarboxylic acid cycle, typical of aerobic bacteria metabolism. This cycle is coupled with
the electron transport chain. It uses O_2 as the terminal electron acceptor and is
referred to as oxidative phosphorylation. This allows ATP synthesis, the aerobic
bacteria energy source, and uses the main part of consumed O_2. ATP synthesis, O_2
consumption, and oxidation of the substrate to CO_2 are thus closely related. It has

been proven from theoretical calculations and observations [4] that the nature of the substrate has a great influence on the ATP requirement for transformation into biomass. Compounds such as pyruvate or acetate need more ATP than glucose, so that more CO_2, and then less biomass, is produced with the same amount of O_2 [4].

In the case of PCL, examination of the biomass amount data of Table 1 leads to the following conclusions. With the pure culture, more carbon is converted to biomass in the case of PCL37000 than with PCL4000, indicating that more ATP is required for PCL4000 biodegradation, perhaps because of its higher crystallinity. Methlylated PCL4000 biodegradation likewise requires more ATP. The actinomycete is more efficient in carbon conversion than the consortium of microorganisms from the compost, although yeast extract is provided in it.

CONCLUSION

Nearly complete biodegradation of PCL4000 and PCL37000 has been observed with microorganisms from an industrial compost. Small amounts of oligomers have been noticed in all cases. Initial oxygen uptake rates are greater for hydroxy- and methoxy-terminated PCL4000 than for PCL37000. With microorganisms from the pure culture, carbon conversion to biomass is slightly higher for PCL37000 than for hydroxy-terminated PCL4000, but is significantly lower than both of them for methoxy-terminated PCL4000.

ACKNOWLEDGMENT

We are very grateful to Mrs. C. Vander Wauven for efficient collaboration and helpful discussions.

REFERENCES

[1]   F. Lefebvre, C. Vander Wauven, and C. David, *Polym. Degrad. Stab., 45,* 347 (1994).
[2]   F. B. Khambutter, F. Warner, T. Russell, and R. S. Stein, *J. Polym. Sci., Polym. Phys. Ed., 14,* 1391 (1976).
[3]   R. D. Swisher, *Surfactant Biodegradation,* Dekker, New York, 1970, p. 444.
[4]   A. H. Stouthamer and H. W. Van Verseveld, in *Comprehensive Biotechnology* (M. Moo-Young, Ed.), Pergamon Press, Oxford, 1985, p. 215.

.

DEGRADABILITY OF POLY(β-HYDROXYBUTYRATE)S. CORRELATION WITH CHEMICAL MICROSTRUCTURE

PIOTR KURCOK, MAREK KOWALCZUK, GRAŻYNA ADAMUS, and ZBIGNIEW JEDLIŃSKI

Institute of Polymer Chemistry
Polish Academy of Sciences
41-800 Zabrze, Poland

ROBERT W. LENZ

Polymer Science and Engineering Department
University of Massachusetts
Amherst, Massachusetts 01003, USA

ABSTRACT

Poly(β-hydroxybutyrate)s (PHB) of different microstructures were synthesized via anionic polymerization of β-butyrolactone initiated by two initiators: 1) supramolecular complexes of alkali metals with asymmetric induction agents, and 2) alkali metal alkoxides. The relationships between chemical microstructure and hydrolytic as well as thermal degradation properties of synthetic and natural PHB are discussed.

INTRODUCTION

Ring-opening polymerization of β-butyrolactone leads to poly(β-hydroxybutyrate)s of different tacticities which mimic those found in nature [1, 2]. Supramolecular complexes of alkali metals, alkali metal naphthalenides, and alkoxides are effective initiators for the anionic polymerization of racemic β-butyrolactone, yielding biodegradable amorphous polyesters [3–5]. Such polymers can be applied as biodegradable plasticizers [6] and for chemical modification of polymers via block

283

and graft copolymerization [7, 8]. It has also been recently demonstrated that polymerization of racemic (R,S) β-butyrolactone initiated with potassium/18-crown-6 supramolecular complex and carried out in the presence of (+) dimethyl tartarate leads to polymer containing predominantly syndiotactic sequences [9].

In this work we report on the relationships between chemical microstructure and hydrolytic as well as thermal degradation properties of synthetic and natural PHB.

RESULTS AND DISCUSSION

Polymer Synthesis and Microstructure

Chemically synthesized poly(β-hydroxybutyrate)s were obtained by ring-opening anionic polymerization of β-butyrolactone. The polymerization of racemic (R,S) β-butyrolactone (β-BL) with potassium tert-butoxide/18-crown-6 complex leads to amorphous polymer. On the other hand, semicrystalline polymer was obtained via polymerization of racemic β-BL initiated with potassium metal/18-crown-6 complexes in the presence of (+) dimethyl tartarate (DMT). The polymerizations were carried out in THF or in bulk.

The molecular weight, polydispersity index, and thermal properties of the chemically synthesized poly(β-hydroxybutyrate)s and that of microbial PHB studied are summarized in Table 1.

The stereoregularity of the investigated polymers was determined by ^{13}C NMR. The results of diad and triad stereosequence distribution analysis, based on the intensity of the split signals of carbonyl carbon C^1 (diad effect [10]) and methylene carbon C^2 (triad effect [1]) depicted in Fig. 1, are summarized in Table 2.

The atactic structure was demonstrated for PHB obtained in the presence of potassium tert-butoxide/18-crown-6 complex (Sample 1). The predominantly syndiotactic structure was revealed for polymer synthesized using potassium/18-crown-6 supramolecular complex in the presence of (+) dimethyl tartarate (Sample 2), while that of microbial PHB is completely isotactic (Sample 3).

TABLE 1. Properties of Poly(β-Hydroxybutyrate) Samples Used for Degradation

Sample[a]	M_n	M_w/M_n	T_g,[b] in °C	T_m,[b] in °C	T_d,[c] in °C
1	20,300	1.15	1	—	255
2	10,600	1.08	3	76	255
3	10,500	2.90	4	178	260

[a]Sample 1 was obtained via anionic polymerization initiated with potassium tert-butoxide/18-crown-6. Sample 2 was obtained via polymerization initiated with potassium supramolecular complex with 18-crown-6 in the presence of DMT. Sample 3 is natural origin PHB.
[b]Scan rate 20°/min.
[c]Maximum decomposition temperature.

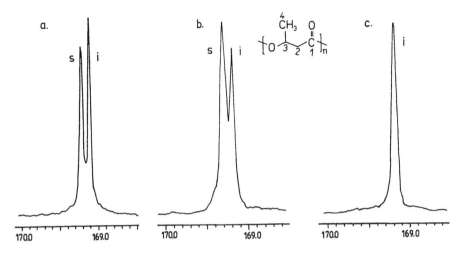

FIG. 1. ^{13}C-NMR spectra (carbonyl region) of (a) atactic PHB, (b) predominantly syndiotactic PHB, and (c) isotactic (natural) PHB.

Hydrolytic Degradation

The hydrolytic degradation of polyesters studied has been carried out in a phosphate buffer (pH 7.4) at a temperature of 70°C. The value of M_w/M_n of the samples studied was observed to broaden slightly during hydrolytic degradation. The molecular weight (M_n) changes are presented in Fig. 2.

It has been revealed that the simple hydrolysis of the polyesters investigated proceeds regardless of their microstructure, similarly to what was described previously by Doi for isotactic PHB and its copolymers of natural origin [11]. Two stages of degradation were observed: the first related to the random hydrolytic chain scission of the ester groups, and then (after a 50% loss of the initial M_n) the onset of molecular weight loss was observed. However, the influence of the sample's crystallinity on the rate of hydrolytic degradation was observed. The sample of atactic PHB degrades faster than the semicrystalline ones.

TABLE 2.   Results of Diad and Triad Stereosequence Distribution Analysis of PHB

| | Tacticity | | | | | |
| | C^1 | | C^2 | | | |
Sample[a]	s_d	i_d	S	H_s	H_i	I
1	50	50	25	24	26	25
2	64	36	41	20	21	18
3	0	100	0	0	0	100

[a]See Footnote [a] in Table 1.

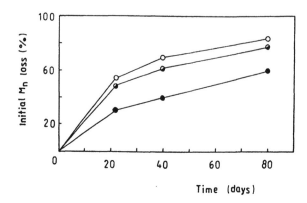

FIG. 2.   Molecular weight (M_n) loss of PHB samples of various tacticities (○) atactic (M_n = 20,300), (◑) predominantly syndiotactic (M_n = 10,600), and (●) isotactic (M_n = 10,500).

Thermal Degradation

The thermal decomposition temperature of the polyesters studied was found to be in the 255 to 260°C region (Table 1). The structure of the products formed during the thermal decomposition of the chemically synthesized polymers has been studied by direct pyrolysis mass spectrometry (DPMS) [12] using a Finnigan MAT SSQ 700 spectrometer. CI-MS analysis of the oligomers formed during pyrolysis conducted on the MS spectrometer solid probe revealed the formation of oligomers with crotonate and carboxylic end groups regardless of PHB tacticity (Fig. 3).

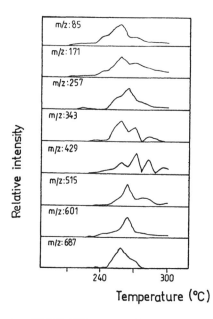

FIG. 3.   Negative ion CI-MS SIC curves of the direct pyrolysis products of atactic PHB.

Similar oligomers were detected recently by Montaudo et al. during pyrolysis of isotactic PHB using the FAB MS technique [13]. These results indicate that regardless of the tacticity of the polyesters studied, the thermal decomposition of poly(β-hydroxybutyrate) proceeds via random scission of polymer chains as proposed previously for natural PHB [14].

CONCLUSIONS

It has been revealed that the chemical microstructure has no substantial influence on the mechanism of both hydrolytic and thermal degradation of atactic, predominantly syndiotactic, and isotactic poly(β-hydroxybutyrate)s. The hydrolytic and thermal degradation proceeds, regardless of the microstructure of the polymers studied, via random scission of the polyester chain. However, the degree of crystallinity of the PHB influences the rate of its hydrolytic degradation.

ACKNOWLEDGMENT

The financial support by the NSF under the framework of the US-Polish M. Sklodowska-Curie Joint Fund II, Grant PAN/NSF-91-60, is acknowledged.

REFERENCES

[1] J. E. Kemnitzer, S. P. McCarthy, and R. A. Gross, *Macromolecules, 26,* 1221 (1993).
[2] Z. Jedliński, in *Handbook of Polymer Synthesis,* Part A (H. R. Kricheldorf, Ed.), Dekker, New York, 1992, p. 645.
[3] Z. Jedliński, P. Kurcok, and M. Kowalczuk, *Macromolecules, 18,* 2679 (1985).
[4] Z. Jedliński, M. Kowalczuk, W. Główkowski, J. Grobelny, and M. Szwarc, *Ibid., 24,* 349 (1991).
[5] P. Kurcok, M. Kowalczuk, K. Hennek, and Z. Jedliński, *Ibid., 25,* 2017 (1992).
[6] Y. Kumagai and Y. Doi, *Makromol. Chem. Rapid Commun., 13,* 179 (1992).
[7] Z. Jedliński, M. Kowalczuk, P. Kurcok, L. Brzoskowska, and J. Franek, *Makromol. Chem., 188,* 1575 (1987).
[8] M. Kowalczuk, G. Adamus, and Z. Jedliński, *Macromolecules, 27,* 572 (1994).
[9] Z. Jedliński, M. Kowalczuk, P. Kurcok, and M. Sokół, *Proceedings of International Conference "Frontiers in Polymerization,"* Liege, 1993, p. 67.
[10] Z. Jedliński, P. Kurcok, M. Kowalczuk, and J. Kasperczyk, *Makromol. Chem., 187* 1651 (1986).

[11] Y. Doi, *Microbial Polyesters,* VCH Publishers, New York, 1990.
[12] G. Montaudo, C. Puglisi, R. Rapisardi, and F. Samperi, *Polym. Degrad. Stab., 31,* 229 (1991).
[13] A. Balisteri, G. Montaudo, M. Giuffrida, R. W. Lenz, Y. Kim, and R. C. Fuller, *Macromolecules, 25,* 1845 (1992).
[14] M. Kunioka and Y. Doi, *Ibid., 23,* 1933 (1990).

EFFECT OF DEGRADATION ON THE MECHANICAL PROPERTIES OF MULTIPHASE POLYMER BLENDS: PHBV/PLLA

S. IANNACE and L. AMBROSIO

Department of Materials and Production Engineering and
    Institute of Composite Materials Technology
University of Naples "Federico II"
Naples, Italy

S. J. HUANG

Institute of Materials Science
University of Connecticut
Storrs, Connecticut, USA

L. NICOLAIS

Department of Materials and Production Engineering and
    Institute of Composite Materials Technology
University of Naples "Federico II"
Naples, Italy

ABSTRACT

The effect of water hydrolysis on the mechanical properties of PHBV/PLLA blends was investigated. The results were interpreted with models able to predict the Young's modulus of multiphase systems. On the basis of the experimental results relative to the pure components, the models were used to calculate the expected values of the blends; these data were therefore compared with the experimental results in order to verify the theoretical predictions. The results show that the modulus of the degraded blends containing a low amount of PLLA is included in a range individuated from the upper and lower bound models and is close to the Halpin–Tsai prediction. This behavior is not present when PLLA is the continuous phase, and this can be due to the reduction of the degradation kinetics of the PLLA phase when PHBV is added to the polymer and forms a partially miscible system. The strength, which depends on both molecular weight and surface erosion that determine crack initiation, decreases faster than the modulus for those systems whose degradation is accompanied with remarkable surface modifications.

INTRODUCTION

Microbial polyesters such as polyhydroxybutyrate and its copolymers with hydroxyvalerate are an interesting class of degradable polymers [1] which is receiving increasing attention in the biomedical field. Because of the absence of specific enzymes, the in-vivo degradation of these materials is claimed to occur by chemical hydrolysis due to the presence of water molecules in the body fluids [1]. Some authors have reported evidence of degradation [2–10] while others have described PHBV as nondegradable [11]. Recent work [12] showed good tolerance of PHBV copolymers inplanted in rats with a degradation of about 40% after 6 months; the reduction of molecular weight was faster for systems with a high content of hydroxyvalerate.

Poly-L-lactide is a component of another interesting family of degradable materials, such as the poly(α-hydroxyacid)s used for resorbable devices and implants [13–18]. Because of their good biocompatibility, they are useful as versatile materials for many temporary therapeutic applications such as osteosynthesis, bone reconstruction, drug delivery, and suture materials [13].

In previous work [19, 20] we showed that solvent-cast films of PHBV and PLLA form heterogeneous systems whose mechanical properties can be modulated through their composition. In the present paper we analyze the effect of the chemical degradation of these blends on their mechanical properties.

MATERIALS AND METHODS

Preparation of Blends and Degradation

PHBV copolymer with 20% HV, under the trade name Biopol, purchased from Malborough Biopolymers Ltd., UK, was used in this study. PLLA was supplied by Novamont S.p.A. (Novara, Italy). Thin films of PHBV/PLLA blends with weight ratios of 100/0, 80/20, 60/40, 40/60, 20/80, and 0/100 were prepared by casting from a solution of chloroform at room temperature. The films were then kept under vacuum overnight at 50°C to allow for the evaporation of the solvent. Films were finally thermally treated for 1 hour at 100°C.

The blends were chemically degraded in phosphate buffer solution at pH 7.4 at 37°C.

Mechanical Properties

Mechanical properties of the samples were studied with an Instron 4202 on dumbbell-shaped samples according to ASTM D 638.

RESULTS AND DISCUSSION

Figures 1 and 2 reported the stress–strain curves for different degradation times for the pure components and for the blends PHBV/PLLA 80/20 and 20/80. In these figures the behavior of the pure components is reported together with that relative to the blends whose continuous phase is constituted by the same parent polymers.

From Fig. 1(a), for pure PLLA, it is possible to observe the reduction of mechanical performance as the consequence of a reduction of the maximum strength and the maximum elongation at break. PLLA shows a yield phenomenon which is followed by a further increase of stress before rupture. As the degradation phenomenon proceeds, this region is progressively reduced until it totally disappears.

This behavior can be understood by analyzing the structure of this polymer and how it changes during degradation. The crystalline regions contain highly ordered sequences of chains which are connected by amorphous zones. The latter are formed by chains that belong to more than one crystalline region and by terminal parts of the chains. Since the first stage of degradation proceeds preferentially in

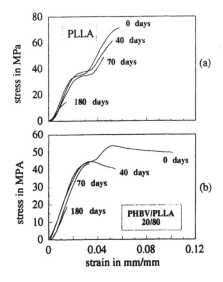

FIG. 1.   Stress–strain curves at different times of degradation for PLLA (a) and the blend PHBV/PLLA 20/80 (b).

the amorphous regions, the stress which is transferred between ordered crystalline regions decreases, and this might result in a progressive reduction of Young's modulus. As a matter of fact, hydrolysis of the amorphous region is usually accompanied by an increase of crystallinity [21, 22], and for this reason the variations in Young's modulus are not easily predicted.

The gradual disappearance of the sigmoidal shape and especially of plastic flow after yield is related to degradation of the amorphous regions. The molecular chains are degraded into small fragments, and since the number of entanglements is reduced, it follows that plastic flow is also reduced. When hydrolysis of the amorphous part is complete, then plastic flow totally disappears and the curves do not present the characteristic sigmoidal shape anymore. Similar results were observed for polyglicolic acid [23].

An analogous behavior is observed for the blend containing 20% PHBV, where PLLA constitutes the continuous phase (Fig. 1b). Here the plastic flow is not characterized by a progressive increase of the stress as was observed for pure PLLA. However, chemical degradation of the amorphous phase progressively induces a reduction of plastic flow, which results in a decrease of the stress and the elongation at break. When the amorphous region is totally degraded, then no more plastic deformation is observed and the samples show brittle fracture with a low level of deformability.

Stress–strain curves of PHBV and of the PHBV/PLLA 80/20 blend were very different from the previous systems because the polyhydroxyalcanoate is above its glass transition temperature at room conditions. The stress–strain curves show a plastic flow region without any sigmoidal shape. This behavior is typical of rubbery materials; they show a high deformability which, in this case, is as high as 15%.

The degradation of the amorphous chains of pure PHBV progressively reduces the deformation of the plastic region (Fig. 2a), and the same trend is also present

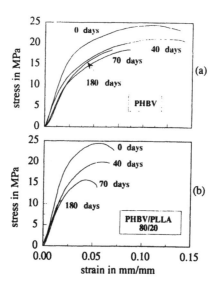

FIG. 2.   Stress–strain curves at different times of degradation for PHBV (a) and the blend PHBV/PLLA 80/20 (b).

for the blend containing 20% PLLA as the dispersed phase (Fig. 2b); here the presence of the faster degradable PLLA contributes to speeding up the reduction of the mechanical performances, as we will discuss later.

The behavior of the different systems investigated are compared in Fig. 3. Here Young's modulus of the blends and the pure components are reported as relative values calculated as the ratio of the actual and the initial value of the nondegraded materials. Young's modulus of pure PLLA and the PHBV/PLLA 20/ 80 blend decrease very slowly at the beginning, and the decay of this mechanical characteristic becomes faster when most of the amorphous region is degraded. On the other hand, systems containing PHBV as the continuous phase show an initial decrease due to the plasticization effect followed by a constant value of the modulus versus time. From Fig. 3 it can be pointed out that the decreasing trend of Young's modulus of the blends is comparable to that of the pure components which constitute the matrix. This effect is more evident if Young's modulus of the blends is reported versus the composition of the system at each degradation time (Fig. 4). The modulus of the degraded blends is included in a range individuated by the upper and lower bound models, and they are close to those predicted by the semiempirical Halpin-Tsai equation [24].

From a practical point of view, this approach can be considered to be a good way to predict mechanical behavior during degradation of the blends at small deformation if the decreasing trend of the pure components is known. On the other hand, information given by the mechanical properties does not clearly demonstrate if there are any effects of the degradation of one phase on the mechanism of degradation of the other phase present in the blend. As a matter of fact, it is reasonable to expect that, due to the partial miscibility of the low molecular weight fractions of the polymers [20], the presence of PHBV can remove the acid fragments that result from the degradation of the PLLA chains, thus contributing to avoidance of the autocatalytic effect of the degradation of PLLA due to a local increase of pH [21, 22]. The discrepancy between the experimental data and the theoretical predictions in Fig. 4, observed for the blend containing 20% PHBV, is an indication that the degradation of the PLLA matrix is affected by the presence

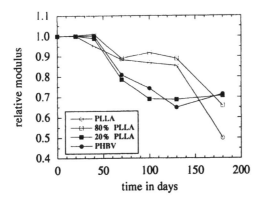

FIG. 3.   Young's modulus versus time of degradation. The values are calculated as the ratio between the actual values and those relative to the systems before the degradation.

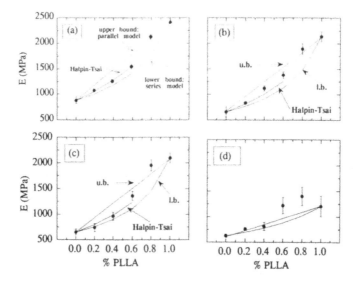

FIG. 4.   Young's modulus versus the weight fraction of PLLA in the blend at different times of degradation: (a) before degradation, (b) 70 days, (c) 100 days, (d) 180 days. The experimental data are compared to the expected values calculated by the Halpin–Tsai equation, the upper and the lower bounds.

of microbial polyester in the structure. More investigations, including molecular weight measurements, are in progress with the purpose of better understanding this behavior.

The relative strength versus time for the blends is reported in Fig. 5. The decreasing trend of multicomponent materials in this case is comparable to that of the pure components which constitute the matrix of that particular blend. The lower values observed for the PHBV/PLLA 80/20 blend with respect to those of pure PHBV depend upon the surface erosion that takes place in all blends containing

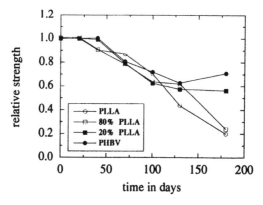

FIG. 5.   Maximum stress versus time of degradation. The values are calculated as the ratio between the actual values and those relative to the systems before the degradation.

PLLA and is absent for pure PHBV. The strength, which depends on both molecular weight and surface erosion that determine crack initiation, decreases faster than the modulus. This behavior was observed for those systems whose degradation is accompanied by remarkable surface modifications.

CONCLUSIONS

The effect of degradation on the mechanical properties of a multiphase system, such as the PHBV/PLLA blend, can be interpreted on the basis of experimental results relative to the pure components. This is particularly true for properties at small deformation such as the elastic modulus. The strength, which depends on both molecular weight and surface erosion that determine crack initiation, decreases faster than the modulus for those systems whose degradation is accompanied with remarkable surface modifications.

A reasonable hypothesis is that the presence of two materials in very close domains, and the partial miscibility of low molecular weight fragments, affect the degradation of both materials: PLLA induces a faster degradation on PHBV chains and, at the same time, PHBV delays the degradation of PLLA molecules.

REFERENCES

[1] Y. Doi, *Microbial Polyesters*, VCH Publishers, New York, 1990.
[2] T. Saito, K. Tomita, and K. Ooba, *Biomaterials, 12*, 309 (1991).
[3] R. L. Kronental, in *Polymer in Medicine and Surgery* (Polymer Science and Technology, Vol. 8, R. K. Kronental, Z. Oser, and E. Martin, Eds.), Plenum Press, New York, 1973.
[4] S. J. Holland, A. M. Jolly, M. Yasin, and B. J. Tighe, *Biomaterials, 8*, 289 (1987).
[5] M. Yasin, S. J. Holland, and B. J. Tighe, *Ibid., 11*, 451 (1990).
[6] S. J. Holland, M. Yasin, and B. J. Tighe, *Ibid., 11*, 206 (1990).
[7] J. C. Knowles and G. W. Hastings, *Ibid., 12*, 210 (1991).
[8] C. Doyle, E. T. Tanner, and W. Bonfield, *Ibid., 12*, 841 (1991).
[9] J. E. Kemnitzer, S. P. McCarty, and R. A. Gross, *Polym. Mater. Sci. Eng., 66*, 408 (1992).
[10] H. Nishida and T. Tokiwa, *J. Appl. Polym. Sci., 46*, 1467 (1992).
[11] N. D. Miller and D. F. Williams, *Biomaterials, 8*, 129 (1987).
[12] S. Gogolewski, M. Jovanovic, S. M. Perren, J. D. Dillon, and M. K. Hughes, *Biomed. Mater. Res., 27*, 1135 (1993).
[13] M. Vert, S. M. Li, G. Spenlehauer, and P. Guerin, *Mater. Sci., Mater. Med., 3*, 432 (1992).
[14] J. M. Brady, D. E. Cutright, R. A. Miller, G. C. Battistone, and E. E. Hunsuck, *J. Biomed. Mater. Res., 7*, 155 (1973).
[15] T. Nakamura, S. Hitomi, S. Watanabe, Y. Shimizu, K. Jamshidi, S. H. Hyon, and Y. Ikada, *J. Biomed. Mater. Res., 23*, 1115 (1989).

[16] J. M. Schakenrad, P. Nieuwenhuis, M. J. Hardonk, I. Molenaar, J. Helder, P. J. Dijkstra, and J. Feijen, *J. Biomed. Mater. Res., 23*, 1271 (1989).

[17] R. R. M. Bos, F. R. Rozema, G. Boering, A. J. Nijenhuis, A. J. Pennings, A. B. Verwey, P. Nieuwenhuis, and H. W. B. Jansen, *Biomaterials, 12*, 32 (1991).

[18] Y. Matsusue, T. Yamamuro, M. Oka, Y. Shikinami, S. H. Hyon, and Y. Ikada, *J. Biomed. Mater. Res., 26*, 1553 (1992).

[19] S. Iannace, H. D. Hwu, S. J. Huang, and L. Nicolais, *Proceeding of 4th World Biomaterial Congress,* Berlin, April 24–28, 1992.

[20] S. Iannace, L. Ambrosio, S. J. Huang, and L. Nicolais, *J. Appl. Polym. Sci.,* Submitted.

[21] S. M. Li, H. Garreau, and M. Vert, *J. Mater. Sci., Mater. Med., 1*, 123 (1990).

[22] S. M. Li, H. Garreau, and M. Vert, *Ibid., 1*, 131 (1990).

[23] C. C. Chu, *J. Appl. Polym. Sci., 26*, 1727 (1981).

[24] R. A. Dickie, in *Polymer Blends,* Vol. 1 (D. R. Paul and S. Newmann, Eds.), Academic Press, New York, 1978, Ch. 8.

ENZYMATIC DEGRADABILITY OF POLY(β-HYDROXYBUTYRATE) AS A FUNCTION OF TACTICITY

PHILIPPA J. HOCKING

Department of Chemistry
McGill University
3420 University Street, Montreal, Quebec H3A 2A7, Canada

MARK R. TIMMINS and THOMAS M. SCHERER

Department of Polymer Science and Engineering

R. CLINTON FULLER

Department of Biochemistry

ROBERT W. LENZ

Department of Polymer Science and Engineering

University of Massachusetts
Amherst, Massachusetts 01003, USA

ROBERT H. MARCHESSAULT

Department of Chemistry
McGill University
3420 University Street, Montreal, Quebec H3A 2A7, Canada

ABSTRACT

The enzymatic degradation of films of synthetic poly[(R,S)-β-hydroxybutyrate], PHB, of various tacticities was investigated by weight loss measurement. Extracellular PHB depolymerases of the bacterium *Pseudomonas lemoignei* and the fungus *Aspergillus fumigatus* M2A were used. In both enzyme systems the nearly atactic samples showed the greatest weight loss; this maximum occurred at a slightly lower percent isotacticity in the *P. lemoignei* system than in the *A. fumigatus*. The maximum degradation rate for the *P. lemoignei* system was double that of the *A. fumigatus*; natural PHB degraded an order of magnitude more rapidly in both enzyme systems. As isotacticity increased, synthetic PHB showed decreased degradation while syndiotactic samples did not appreciably degrade. Results are interpreted in terms of both crystallinity and stereochemistry; the *A. fumigatus* system is more affected than the *P. lemoignei* by the presence of the *S* repeat units. Total weight loss data suggest that the enzymes are capable of *endo* cleavage.

INTRODUCTION

Poly(β-hydroxybutyrate), PHB, is a biodegradable polyester made by bacteria as a storage material [1–3]. In the natural polymer, all chiral carbon atoms have the *R* configuration, giving perfectly isotactic polymer. Racemic monomer can be polymerized to yield synthetic PHB with a variety of tacticities: stereoblock isotactic [4–8], atactic [9, 10], and syndiotactic [11–13]. Investigating the enzymatic degradability of these stereocopolymers allows greater understanding of the fundamental principles controlling the biodegradation of PHB.

EXPERIMENTAL

The synthetic PHB used in this study was made by ring-opening polymerization of (R,S)-β-butyrolactone using a methylaluminoxane catalyst [11]. It was then fractionated according to tacticity, and solvent-cast into films. Bacterial PHB, obtained from Marlborough Biopolymers (Billingham, UK), was similarly cast. The enzymes used were extracellular PHB depolymerases isolated from the bacterium *Pseudomonas lemoignei* [14] and the fungus *Aspergillus fumigatus* [15]. Degradation was examined by placing the polymer films into buffer containing a fixed activity of enzyme, and monitoring weight loss normalized to initial film surface area [16]. Control experiments included degradation of bacterial PHB and degradation of each material in the absence of enzyme.

RESULTS AND DISCUSSION

The synthetic racemic PHB films varied in tacticity from 34 to 88% isotactic diads, as determined by integration of the two components of the ^{13}C NMR carbonyl peak. Crystallinities of these materials increased to either side of ~50%

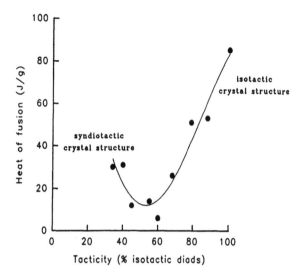

FIG. 1.   Crystallinities of PHB films used, as determined by heats of fusion.

isotactic diads as shown in Fig. 1; isotactic and syndiotactic crystal structures were easily distinguished by wide-angle x-ray diffraction [11].

Degradation of Synthetic PHB Films

Figure 2 shows typical results for the bacterial system, plotted as weight loss (mg/cm^2) versus time. Total weight loss for each material is shown as the percentage of initial sample weight. Results for both enzyme systems are summarized in

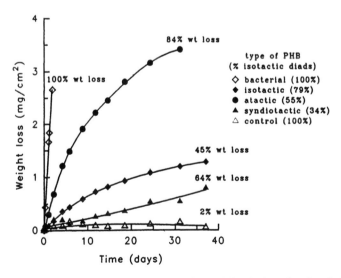

FIG. 2.   Weight loss of natural and synthetic PHB by *P. lemoignei* and *A. fumigatus* extracellular depolymerase.

Fig. 3, where values for all replicates of each sample were averaged and plotted as weight loss after 500 hours of degradation time versus isotactic diad content. Control samples, treated identically to the test samples but with no enzyme added, are also shown.

Examination of these results confirms that both crystallinity and regularity of R configuration are significant factors in degradation behavior. Previous experiments have shown that decreasing crystallinity increases degradability [17–19], but decreasing the length of the R sequence decreases degradability [18, 20, 21]. These two factors act in opposing directions in isotactic racemic PHB, such that an optimal balance is reached at 55–60% isotactic diads; at higher isotacticities, the negative effect of higher crystallinity is greater than the positive effect of higher isotacticity, so degradability decreases. In syndiotactic racemic PHB, crystallinity increases with decreasing isotacticity; as both these factors have a negative effect on degradability, the observed degradation is minimal.

Differences between the two enzyme systems are clearly visible in Fig. 3. The maximum degradation for *A. fumigatus* enzymes is about half that for *P. lemoignei*, despite the fact that both systems degraded bacterial PHB at the same rate. Highly isotactic samples (79–88% isotactic diads) degraded significantly in the *P. lemoignei* enzymes but not in the *A. fumigatus*, and the whole degradation peak occurs at a slightly lower isotacticity for the *P. lemoignei* system. These observations suggest that the *P. lemoignei* enzymes are less hindered by the unnatural *S* unit than are the *A. fumigatus* enzymes.

Total weight loss data can be used to speculate on the enzyme degradation mechanism. Identification of small oligomers as the primary degradation products of bacterial PHB [22–25] suggests that degradation occurs preferentially from the chain ends (*exo* attack). If this were the only mechanism involved, degradation of racemic samples would cease before reaching 50% of the initial sample weight, as

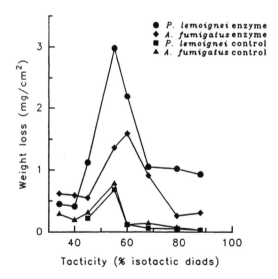

FIG. 3.   Weight loss of synthetic PHB after 500 hours of degradation by *P. lemoignei* and *A. fumigatus* extracellular depolymerases.

the enzymes would not be able to penetrate the blocks of S units which would accumulate at the chain ends, and these undegradable S units comprise 50% of the initial sample mass. However, samples in this study showed total weight losses up to 80% of initial mass, indicating that at least a small amount of *endo* cleavage must occur to provide new, degradable chain ends. That the synthetic racemic samples degrade approximately an order of magnitude less readily than bacterial PHB gives some indication of the relative preference the enzymes have for *exo* cleavage over *endo*.

CONCLUSIONS

This work has shown that depolymerases from both *P. lemoignei* and *A. fumigatus* are capable of degrading synthetic racemic PHB, within the restrictions of crystallinity and stereochemistry. Thus, atactic racemic PHB shows the greatest degradability, highly isotactic racemic PHB degrades less well due to high crystallinity, and syndiotactic racemic PHB shows minimal degradation due to both high crystallinity and low isotacticity. Comparison of the two enzyme systems indicates that the *A. fumigatus* depolymerase is more sensitive than the *P. lemoignei* depolymerase to the stereochemistry of the PHB substrate. Finally, total final weight losses imply that the degradation mechanism includes *endo* attack.

ACKNOWLEDGMENTS

This work was supported by the Natural Sciences and Engineering Research Council of Canada (NSERC), the National Science Foundation of the USA (NSF), and Xerox Corporation. P.J.H. thanks NSERC for scholarship support.

REFERENCES

[1] P. A. Holmes, in *Developments in Crystalline Polymers,* Vol. 2 (D. C. Basset, Ed.), Elsevier, New York, 1989, Chapter 1.
[2] Y. Doi, *Microbial Polyesters,* VCH Publishers, New York, 1990.
[3] P. J. Hocking and R. H. Marchessault, in *Chemistry and Technology of Biodegradable Polymers* (G. J. L. Griffin, Ed.), Blackie A & P, Glasgow, 1994, Chapter 4.
[4] R. A. Gross, Y. Zhang, G. Konrad, and R. W. Lenz, *Macromolecules, 21,* 2657 (1988).
[5] S. Bloembergen, D. A. Holden, T. L. Bluhm, G. K. Hamer, and R. H. Marchessault, *Ibid., 22,* 1656 (1989).
[6] Y. Zhang, R. A. Gross, and R. W. Lenz, *Ibid., 23,* 3206 (1990).
[7] M. Benvenuti and R. W. Lenz, *J. Polym. Sci., Polym. Chem. Ed., 29,* 793 (1991).
[8] A. D. Pajerski and R. W. Lenz, *Makromol. Chem., Macromol. Symp., 37,* 7 (1993).

[9]   K. Teranishi, M. Iida, T. Araki, S. Yamashita, and H. Tani, *Macromolecules, 7*, 421 (1974).

[10] M. Iida, T. Araki, K. Teranishi, and H. Tani, *Ibid., 10*, 275 (1977).

[11] P. J. Hocking and R. H. Marchessault, *Polym. Bull., 30*, 163 (1993).

[12] J. E. Kemnitzer, S. P. McCarthy, and R. A. Gross, *Macromolecules, 26*, 1221 (1993).

[13] J. E. Kemnitzer, S. P. McCarthy, and R. A. Gross, *Ibid., 26*, 6143 (1993).

[14] M. R. Timmins, P. J. Hocking, T. M. Scherer, R. C. Fuller, R. W. Lenz, and R. H. Marchessault, In Preparation.

[15] M. Diglio and S. Goodwin, In Preparation.

[16] P. J. Hocking, M. R. Timmins, T. M. Scherer, R. C. Fuller, R. W. Lenz, and R. H. Marchessault, *Macromol. Rapid Commun., 15*, 447 (1994).

[17] Y. Kumagai, Y. Kanesawa, and Y. Doi, *Makromol. Chem., 193*, 53 (1992).

[18] Y. Doi, Y. Kumagai, N. Tanahashi, and K. Mukai, in *Biodegradable Polymers and Plastics* (M. Vert, J. Feijen, A. Albertsson, G. Scott, and E. Chiellini, Eds.), Royal Society of Chemistry, Spec. Publ. 109, Cambridge, 1992, p. 139.

[19] M. Parikh, R. A. Gross, and S. P. McCarthy, *Polym. Mater. Sci. Eng., 66*, 408 (1992).

[20] J. J. Jesudason, R. H. Marchessault, and T. Saito, *J. Environ. Polym. Degrad., 1*, 89 (1993).

[21] J. E. Kemnitzer, S. P. McCarthy, and R. A. Gross, *Macromolecules, 25*, 5927 (1992).

[22] F. P. Delafield, M. Doudoroff, N. J. Palleroni, C. J. Lusty, and R. Contopoulos, *J. Bacteriol., 90*, 1455 (1965).

[23] C. J. Lusty and M. Doudoroff, *Proc. Natl. Acad. Sci. USA, 56*, 960 (1966).

[24] K. Nakayama, T. Saito, T. Fukui, Y. Shirakura, and K. Tomita, *Biochim. Biophys. Acta, 827*, 63 (1985).

[25] T. Tanio, T. Fukui, Y. Shirakura, T. Saito, K. Tomita, T. Kaiho, and S. Masamune, *Eur. J. Biochem., 124*, 71 (1982).

GENETICALLY MODIFIED STARCH AS AN INDUSTRIAL RAW MATERIAL

CHRISTER JANSSON, SATHISH PUTHIGAE, CHUANXIN SUN, and CARL GRENTHE

Department of Biochemistry
The Arrhenius Laboratories
Stockholm University
S-10691 Stockholm, Sweden

ABSTRACT

Starch is an important raw material for industrial applications, both for food and nonfood purposes. Of particular interest is the use of starch as a nonpetroleum chemical stock for the manufacture of biodegradable polymers. Annual EC starch production is nearing 10 million tons, with 80% from cereals and 20% from potatoes, and grows at 4–5% annually. The potential for genetically modified starch is considered very high. Such starch offers significant advantages: 1) chemical modifications, which are expensive and environmentally hazardous, are replaced; 2) novel carbohydrates can be produced.

STARCH AS AN INDUSTRIAL RAW MATERIAL

Starch is an important raw material for industrial applications, both for food and nonfood purposes, and the market is increasing. Of particular interest is the use of starch as a nonpetroleum chemical feedstock for the manufacture of biodegradable polymers, such as "plastics," and as a noncellulose feedstock in the paper industry. In addition to promising applications outside the food industry, nutritionally and functionally modified starches have great potential as new and improved food additives. The annual starch production within the EC is approaching 10 million tons (MT) and it grows at 4–5% per year with 80% originating from cereals

303

and 20% from potato. Of the consumed starch, 45% is used as native or modified starch in foodstuff, polymers, paper, chemicals, boards, and pharmaceuticals, while the rest is used as starch hydrolysates.

STARCH AND SYNTHESIS OF STARCH

Starch is a mixture of amylose and amylopectin, both glucose polymers. Amylose is a mostly linear polymer of 200–2000 α-1,4-bonded glucose moieties with rare α-1,6 branch points. Amylopectin, on the other hand, is highly α-1,6 branched, with a complex structure of 10^6–10^8 MW and up to 3×10^6 glucose subunits, making it one of the largest biological molecules in nature (Fig. 1; see Ref. 1 for a detailed description of the amylose and amylopectin molecules). In the plant, starch is deposited as starch granules, primarily in amyloplasts of endosperm (seeds), tubers, and roots. Figure 2 shows the central enzymes and the metabolic pathways in starch synthesis (see Refs. 2 and 3 for recent reviews).

In most plants starch consists of 20–30% amylose and 70–80% amylopectin. The structure of the amylose and amylopectin molecules, the amylose/amylopectin ratio, the degree of substitution, and the association of lipids and proteins are responsible for the functional qualities of starch, and thereby affect properties such as gelatinization, retrogradation, viscosity, fermentability, behavior as granules, and behavior as pure and mixed polymer sheets. The size of the starch granules is heterogeneous, with a diameter of 0.5–100 μm. In some cereals, such as barley, starch is composed of a fraction with large granules (A-starch) and a fraction with small granules (B-starch). Of the various forms of modified starches that occur in nature, only phosphorylation of potato starch is understood to some extent.

Concomitant with the increased interest in starch as an industrial raw material is a growing demand for production of modified starch in transgenic plants [4–6].

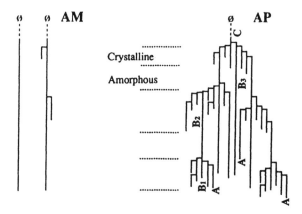

FIG. 1.   The structure of amylose (AM) and amylopectin (AP). The crystalline branch clusters and the amorphous regions between the clusters are shown. The A-chains are unbranched whereas B-chains and the single C-chain with the reducing end ($\oslash$) are branched. The B-chains are further divided according to length from short B1-chains to long B3-chains (see Ref. 1 for references).

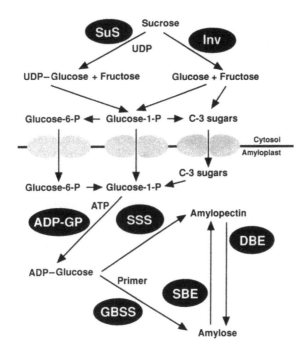

FIG. 2.   Schematic representation of the metabolic pathway for starch synthesis in nonphotosynthetic tissues. SuS = sucrose synthetase, Inv = sucrose invertase, ADP-GP = ADP-glucose pyrophosphorylase, SSS = soluble starch synthase, GBSS = granule-bound starch synthase, SBE = starch branching enzyme, DBE = debranching enzyme.

This approach offers the possibility to replace much of the postharvest chemical modifications, which are environmentally hazardous, expensive, and time-consuming. It also makes possible the production, *in planta*, of novel carbohydrates. A desirable product in transgenic plants would be all- or high-amylose starch for the polymer industry. This could be achieved by inactivation of all *sbe* genes (encoding SBE). Selective inactivation of *sbe* genes would be expected to result in starches with altered branching patterns. Other examples of postharvest modifications that could be replaced by a transgenic approach are production of crosslinked starch for increased stability, production of starch with well-defined granule sizes, and substitutions (hydroxypropylation, methylation, carboxylation, phosphorylation, etc.). Production of novel carbohydrates in transgenic plants has been demonstrated by the synthesis of cyclic dextrines from starch [5].

CEREAL AND POTATO STARCH—A COMPARISON

Cereals are important crop and starch sources. The industry for processing of cereals such as corn, rice, wheat, and barley is at hand and very well developed. The storage and handling characteristics for cereal seeds are excellent, and superior to those of tubers like potato. This suggests that production costs, and thereby shelf prices, for bioengineered cereal starch will be competitive. The annual production of barley is less than that for corn, rice, or wheat; around 167 MT worldwide and 48 MT in the EC. (Corresponding numbers for wheat are 565 and 85 MT, respectively). Over 60% of the barley production in the EC is used in the starch-consuming industry. However, barley is globally an important crop for food, feed, and malt production. Barley is also a sturdy plant that can grow in marginal areas. In addition, barley is a diploid species (as opposed to, e.g., potato and wheat) and genetically flexible. Finally, knowledge in barley genetics and breeding is high, and large collections of barley mutants are available. Particle-gun mediated transformation of cereals, including corn, rice, wheat, and barley, has been reported [7, 8].

The annual production of potato is approximately 260 MT worldwide and 45 MT in the EC. Potato is a very important starch source due to the large yield in tubers. The incorporated phosphate groups in the amylopectin chains contribute to the mild taste of potato starch and to its ability to form clear, transparent solutions and gels, and thus renders it an attractive starch in the food industry. Transformation of potato with *Agrobacterium tumefaciens* is well established, and transgenic potatoes with altered starch composition and yield have been described [5].

CONCLUSION

Production of modified starch in transgenic plants has great future potential, especially given the fact that crude oil and cellulose resources are shrinking, and also taking into account the increasing environmental awareness.

REFERENCES

[1]   W. R. Morrison and J, Karkalas, in *Methods in Plant Biochemistry* (P. M. Dey, Ed.), Academic Press, London, 1990, p. 323.

[2]   J. Preiss, *Oxford Surveys of Plant Molecular and Cell Biology,* Vol. 7, 1994, p. 59.

[3]   P. Sathish, C. Sun, A. Lönneborg, and C. Jansson, *Prog. Bot., 56* (1994), In Press.

[4]   R. G. F. Visser and E. Jacobsen, *Trends Biotechnol., 11*, 63 (1993).

[5]   C. K. Shewmaker and D. M. Stalker, *Plant Physiol., 100*, 1083 (1992).

[6]   G. M. Kishore and C. R. Sommerville, *Curr. Opin. Biotechnol., 4*, 152 (1993).

[7]   P. Christou, *Ibid., 4*, 135 (1993).

[8]   D. McElroy and R. I. S. Bretell, *Trends Biotechnol., 12*, 62 (1994).

SILYLATED CELLULOSE MATERIALS IN DESIGN OF SUPRAMOLECULAR STRUCTURES OF ULTRATHIN CELLULOSE FILMS

DIETER KLEMM and ARMIN STEIN

Institute of Organic and Macromolecular Chemistry
University of Jena
D-07743 Jena, Germany

ABSTRACT

To use the structure-forming potential and the biodegradability of cellulose and nonionic cellulose ethers, we developed synthesis pathways for soluble and regenerable silyl celluloses suitable for the design of advanced materials. A 6-0-silylation of cellulose takes place in a hetero-geneous phase reaction in the presence of ammonia-saturated polar aprotic solvents at $-15°C$ with thexyldimethylchlorosilane. After 2,3-di-O-methylation, this type of regioselectively-substituted cellulose deriva-tives yields sensor matrices for the detection of halohydrocarbons in air. On the other hand, thexyldimethylsilyl celluloses and trimethylsilyl celluloses with degrees of substitution in the 2.6 to 3.0 range form mono- and multilayered supramolecular structures by applying the Langmuir-Blodgett technique and, after desilylation, oriented ultrathin cellulose films.

EXPERIMENTAL

Synthesis of Trimethylsilyl Celluloses

Starting from different types of cellulose, trimethylsilyl celluloses were prepared by silylation with an excess of hexamethyldisilazane at 80°C after dissolution in dimethylacetamide/LiCl [1]. Their characteristics are listed in Table 1.

Synthesis of 6-0-Thexyldimethylsilyl-2,3-di-O-methylcellulose

Avicel (2.0 g) was dried at 105°C in vacuum and added to ammonia-saturated DMF at −15°C. After stirring for 1 hour at −15°C, thexyldimethylchlorosilane (1.5 mol/mol repeating unit) was added dropwise, stirring was continued for 1 hour at this temperature, and for an additional 6 hours up to 60°C. The silylated polymer was precipitated in water at 20°C, collected, and dried. Yield: 96%, DS_{Si} 0.99, DP_w 200. Methyl ether formation was carried out by conversion with iodomethane in the presence of sodium hydride in THF at room temperature [3].

Measurements

^{1}H-NMR and ^{13}C-NMR spectra were recorded with a Bruker-WP 400 spectrometer in toluene-d_8 at 80°C. IR spectra were run on an FT-IR spectrometer 60 SK (Nicolet).

Formation of Films of Silylated Celluloses

Mono- and multilayers of trimethylsilyl celluloses were formed by a Langmuir–Blodgett technique described in Ref. 4. Sensor films of the silylated celluloses were spin coated on Al_2O_3 substrates in contact with Pt electrodes (Inter-Digital-Resistor) for determination of conductivity in the presence of halohydrocarbons in air [5].

TABLE 1. Characteristics of Trimethysilyl Celluloses

Cellulose		Trimethylsilyl cellulose			
Type	DP^a	DP_n^b	DP_w^b	$Si/\%^c$	DS^d
Avicel PH 101 (Fluka)	200	80	220	21.98	2.9 $(2.86)^e$
Spruce sulfite pulp	600	160	1100	21.83	2.9
Cotton linters	1400	420	3100	21.20	2.7

[a]Degree of polymerization, in cuoxam by viscosimetry.
[b]Estimated by GPC in THF.
[c]Determined gravimetrically (SiO_2) [2].
[d]Degree of substitution, from Si/%.
[e]Calculated from ^{1}H-NMR spectra.

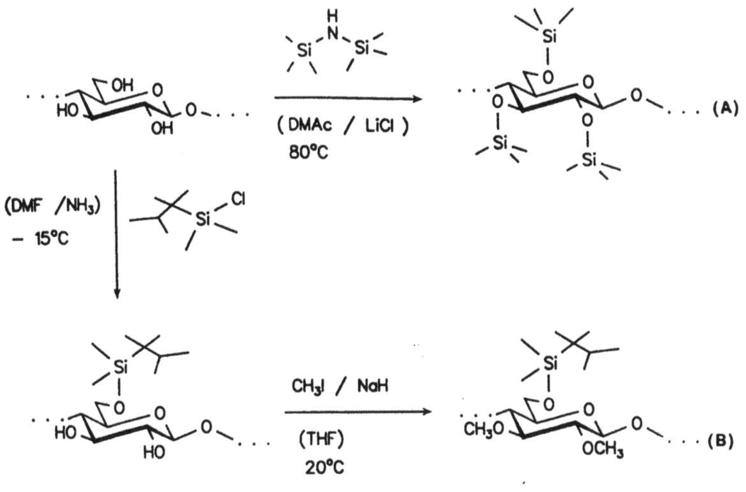

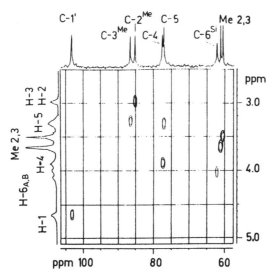

SCHEME 1.

RESULTS AND DISCUSSION

One aim of our present work consists of an investigation of cellulose derivatives with regular molecular structures suitable for building-up supramolecular architectures in biodegradable advanced materials.

Starting from different cellulose samples (see Table 1), trimethylsilyl celluloses were prepared by complete silylation in homogeneous cellulose solutions in dimethylacetamide/lithium chloride (Scheme 1, top). A regioselective 6-O-thexyldi-

FIG. 1. The $^{13}C[^{1}H]$-NMR spectra of 6-0-thexyldimethylsilyl-2,3-di-O-methylcellulose in deuterotoluene.

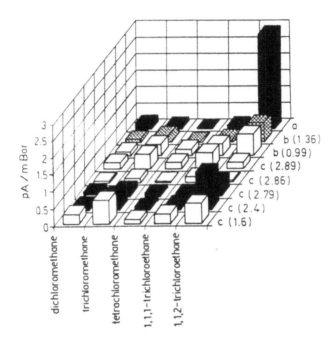

FIG. 2.   Sensor sensibility (pA/mBar) for 6-0-thexyldimethylsilyl-2,3-di-*O*-methylcel-
lulose (a), thexyldimethylsilyl cellulose (b), and trimethylsilyl cellulose (c) [5]. In parentheses: degree of substitution of silyl groups.

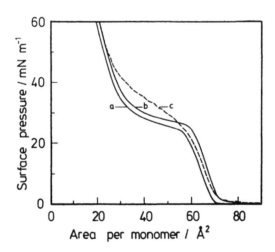

FIG. 3.   The surface pressure area isotherms recorded at 20°C on air–water interfaces
for trimethylsilyl celluloses; DP 200, DS 2.9, spread from chloroform (a); DP 1400, DS 2.7,
from chloroform (b); DP 1400, DS 2.7, from *n*-hexane (c).

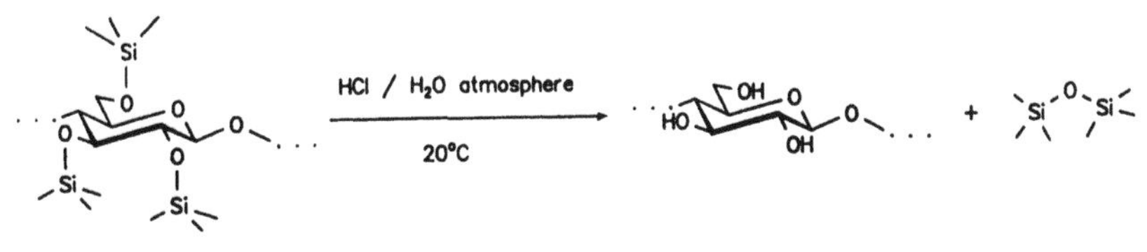

SCHEME 2.

methylsilylation was carried out in ammonia-saturated polar aprotic solvents followed by 2,3-O-methylation (Scheme 1, bottom).

The regular molecular structure of the methylated thexyldimethylsilyl cellulose has been confirmed by $^{13}C[^{1}H]$-NMR spectra (see Fig. 1).

In the course of studies on polymers as sensor materials for the detection of halohydrocarbons in air, the synthesized celluloses have been investigated. As seen from Fig. 2, spin-coated sensor films of thexyldimethylsilylated methyl cellulose on Inter-Digital-Resistors (see Experimental Section) show a significant sensitivity toward 1,1,2-trichloroethane compared with the other silylcelluloses [5].

On the other hand, 6-0-thexyldimethylsilyl cellulose and trimethylsilyl celluloses (degree of substitution 2.6–2.9) have been used as soluble intermediates for building up well-defined mono- and multilayered ultrathin films of regenerated cellulose. Spreading of the silylcelluloses from chloroform or n-hexane solutions and compressing the polymer molecules on an air–water interface by the Langmuir–Blodgett technique [4] form monolayers up to a surface pressure of 24 nM/m (see Fig. 3). At higher surface pressures a plateau region is reached and the monofilm collapses. After transfer of the layer onto hydrophobized glass slides, silicon wafers, or gold surfaces and subsequent in-situ desilylation, mono- and multifilms of cellulose resulted [4]. The desilylation (Scheme 2) can be carried out in a simple way with gaseous HCl with 30 seconds.

In the case of n-octyldimethylsilyl cellulose, a comparable formation of monolayers on the air–water interface can be observed, but the low surface pressure of 10 mN/m does not allow the transfer of these layers onto a substrate. On the other hand, trimethylsilyl celluloses with degrees of substitution lower than 2.5 and 6-0-thexyldimethylsilyl cellulose (without methylation of the secondary OH groups) are unsuitable for preparing monolayers because of molecular aggregation in the solvents used.

CONCLUSIONS

Silylated celluloses are suitably soluble and regenerable polymers to design supramolecular structures in Langmuir–Boldgett and spin-coated films. Trimethylsilyl celluloses with degrees of substitution higher than 2.5 form well-defined mono- and multilayered architectures. In-situ desilylation of these films represents a convenient method to generate oriented thin hydrophilic and biodegradable cellulose films.

ACKNOWLEDGMENTS

The authors would like to acknowledge the financial support of the Bundes-ministerium für Forschung und Technologie. They are grateful to Prof. Wegner and Prof. Wenz (Max Planck Institute for Polymerresearch, Mainz), and Prof. Kossmehl (FU Berlin) for Langmuir–Blodgett and sensor measurements.

REFERENCES

[1] W. Schempp, Th. Krause, U. Seifried, and A. Koura, *Papier, 35*, 547 (1981).
[2] A. Stein and D. Klemm, *Makromol. Chem., Rapid Commun., 9*, 569 (1986).
[3] U. Erler, P. Mischnick, A. Stein, and D. Klemm, *Polym. Bull., 29*, 349 (1992).
[4] M. Schaub, G. Wenz, G. Wegner, A. Stein, and D. Klemm, *Adv. Mater., 5*, 919 (1993).
[5] G. Kossmehl, Unpublished Results.

INDEX

Milton Keynes UK
Ingram Content Group UK Ltd.
UKHW051947071024
449327UK00026B/2203